Moralizing the environment

D0139581

TECHNICAL COLLEGE OF THE LOWCOUNTRY
LEARNING RESOURCES CENTER
POST OFFICE BOX 1288
BEAUFORT, SOUTH CAROLINA 29901-1288

Moralizing the environment
Countryside change, farming and pollution

Philip Lowe
University of Newcastle upon Tyne

Judy Clark
University College London

Susanne Seymour
University of Nottingham

Neil Ward
University of Newcastle upon Tyne

UCL
PRESS

© Philip Lowe, Judy Clark, Susanne Seymour and Neil Ward, 1997
This book is copyright under the Berne Convention.
No reproduction without permission.
All rights reserved.

First published in 1997 by UCL Press
UCL Press Limited
1 Gunpowder Square
London EC4A 3DE
UK

and

1900 Frost Road, Suite 101
Bristol
Pennsylvania 19007–1598
USA

The name of University College London (UCL) is a registered
trade mark used by UCL Press with the consent of the owner.

British Library Cataloguing-in-Publication Data
A catalogue record for this book is available from the British Library.

**Library of Congress Cataloging-in-Publication-Data
are available**

ISBNs: 1–85728–839–4 HB
         1–85728–840–8 PB

Typeset in Garamond by Graphicraft Typesetters Ltd, Hong Kong.
Printed by Arrowhead Books Limited, Reading, UK.

Contents

Contents

Contents

Preface

There was a time when pollution was equated with the urban and the industrial. It was factory chimneys and outfalls and town sewers that belched the grime, slime and smog into the atmosphere, the water system and the natural environment. In contrast, the countryside and agriculture were associated with a natural order of things, with rural communities existing apart from the corrupting influences of urban industrialism, and farmers working in harmony with nature to produce our food.

But things have changed. What were previously mutually exclusive categories of "agriculture" and "pollution" have been brought together in a new, morally charged atmosphere. In Britain in the 1980s, amid the political turmoil surrounding the privatization of the water industry, new categories of environmental risk emerged. The spillage of farm effluents into rivers and streams shifted from being a little regarded technical side-effect of efficient agricultural production, to being an example of wrongdoing, a breach of a new environmental morality. Farmers, once respected as the guardians of the countryside, now stood indicted of ecological crime.

Moralizing the environment is a study of how this shift came about. It examines the emergence of the farm pollution problem in Britain in the 1980s. It draws upon a study of the regulation of farm wastes – cattle slurry, silage effluent and the dirty water from farmyards – conducted between 1989 and 1995. Detailed surveys and ethnographic fieldwork were carried out in the south-west of England among dairy farmers, pollution inspectors, agricultural advisers and environmentalists. In trying to get to grips with farm pollution they were pursuing different notions not only of sound agricultural practice but also of nature, morality and the law. What ultimately was at stake was who could be trusted to safeguard the countryside.

The study operates at three levels. At one level, it is a study of a change in regulatory regimes. Agriculture has passed from being one of the least to one of the most formally regulated sectors from the point of view of pollution control, and the book examines what this entails for regulatory and farming practice. In effect, what this shift has meant is the insertion into the agricultural community at the local level of Pollution Inspectors armed with new legal powers and under public and political pressure to take a tough stance over farm pollution. At another level, therefore, the study is one of boundary maintenance and change, revealing how new powers and priorities are negotiated with a particularly entrenched and traditional occupational community. But these developments are linked to wider transitions in social meanings about what is natural and what is unnatural. At its broadest, therefore, the study is one of the eclipse of farming as a source of natural values by the new environmental morality and how this shift is related to a changing rural world.

Acknowledgements

We would like to thank the Economic and Social Research Council for funding this study under its Joint Agriculture and Environment Programme (Grant W 103 25 1008); Graham Cox, David Goodman, Richard Munton and Mike Winter for their help in the development and management of the research; Graham Cox, Jonathan Murdoch and Michael Mayerfield Bell for their comments on a draft of this book; Pam and David Rosenthall and Alastair Shaw for their help in arranging the farm survey fieldwork; and the Devon dairy farmers, farming officials, environmental activists, civil servants of the Ministry of Agriculture, Fisheries and Food and the Department of the Environment and staff of the National Rivers Authority and Agricultural Development and Advisory Service for their time and patience in being interviewed about their experiences and "shadowed" in their work.

CHAPTER ONE

Moralizing the environment: understanding farm pollution

Introduction

Imagine the scene. A gentle river runs between small dairy farms tucked into the folds of a quiet Devon valley. The water is foaming, there is a stench of ammonia in the air and dead fish float on the surface. Someone's slurry store has overflowed. This beautiful countryside has been desecrated, its natural waters have been polluted and its wildlife has suffered. The farmer responsible could be prosecuted and, if so, may face a heavy fine. But he too feels he is a victim. Persistent rain led to the overflow and if the rain continues it will eventually cleanse the river. After all, specialized dairy farming has for more than a century shaped and maintained the countryside in these parts. Of course, it has had to modernize and intensify to stay in business, but government has encouraged it to do so, and the public take for granted the ready supply of milk produced. One consequence of the technological revolution in dairy farming, though, has been a much greater volume of effluents to be handled, which involves farmers in additional costs, as well as in tasks that are unpleasant and often hazardous. What causes them to feel beleaguered is a growing climate of intolerance they experience if, by accident, these wastes reach the rivers and streams that cross their farms.

An unfortunate but largely unavoidable side-effect of vital production activities, or the desecration of the environment by careless and wasteful practices? These are two starkly contrasting views of pollution. They mark a basic perceptual divide between seeing pollution as a technical side-effect of production and seeing it as something discreditable that attracts blame.

The crucial distinction is whether or not any moral opprobrium is attached, whether it is wrong or simply unfortunate to pollute.

In the past there has certainly been a rhetoric – encapsulated in the Victorian saying "where there's muck, there's brass" – which has associated grime with graft, pollution with industriousness. On the other hand, within Victorian society, there was a counter-reaction to the abominable conditions of the industrializing towns and cities, which represented industrialization as a corrupter of the moral and social order, as well as of human health, traditional values and the physical environment. These anti-industrial values were embodied in institutions and campaigns to preserve aspects of traditional culture and the natural world from the ravages of urban–industrial growth. Nature and the countryside came to provide an alternative value system, one that stood as an indictment of the bleakness and degradation of the industrial age.

By the early twentieth century, rural preservation had become a significant force in British politics, dedicated to protecting nature and the countryside from industrial advance. The need to protect agriculture and create conditions in which it could flourish was part of the conventional wisdom of rural preservation. The preservationists tended to have a highly romantic and idealized view of farming, summed up by G. M. Trevelyan in his *English Social History* (1942) in the following terms: "Agriculture is not merely one industry among many, but is a way of life, unique and irreplaceable in its human and spiritual values." Agriculture had been in a chronic state of depression since the 1880s, which contributed to its image as an occupation steeped in tradition and at one with the natural world. Farming practices seemed to pose no possible threat to other rural interests and pursuits. On the contrary, it was felt that the debilitated condition of farming exacerbated many other threats to the countryside, such as urban encroachment and the decline of rural communities. A secure and revitalized agriculture was seen as the essential conserver of both the social life and the natural beauty of the countryside.

At the same time, social regulation[1] gradually emerged to curb the worst dangers and excesses of industrial advance, championed largely by urban social reformers, the labour movement and professional and scientific groups. The German sociologist Ulrich Beck (1992a,b) has pointed out that part of the constitution of industrial society is a system of rules for dealing with industrially produced hazards and insecurities, which contains and redistributes their consequences. These rules include procedures such as insurance contracts, compensation agreements, liability law, regulatory systems, preventive measures and after-care provisions. Underpinning them is a calculus of risks in which statistics, for example on mortality rates, take the place of moral imperatives, thereby allowing "technological moralization" or a "type of ethics without morality" (Beck 1992b: 99). In this way, industrial society has sought to come to grips with the uncertainty and hazards it systematically creates.

Thus, in the past, pollution could come to be regarded in relative and not absolute terms. Considered in the context of urban-based production, it was seen as a necessary concomitant of jobs and prosperity, as an unfortunate but unavoidable counterpart of industriousness. Formal efforts to curb the worst excesses of industrial pollution emerged pragmatically in Britain in response to specific problems and were informed by the approach of "best practicable means". Quite separately, factory legislation and public health legislation sought to provide basic protection for workers and residents. Implicit within such measures and the way they were implemented was a trade-off between the benefits and risks from the local industries on which particular districts depended.

Only with the advent of the contemporary environmental movement was the equation between pollution and prosperity systematically challenged. One of the most important achievements of the environmental movement has been to establish an abstracted conception of pollution (or pollutants). Environmental groups have been assisted in this by social change. Fewer people directly depend on polluting industries, and the expanding geographical impact of pollution has affected groups and areas not dependent on the particular polluting industries. Widely publicized investigations into the effects of two particular types of pollutant – radioactive fallout from nuclear tests and the residues of persistent organochloride insecticides – catalyzed profound shifts in popular attitudes (see, for example, Commoner 1966). Alarming evidence of the universal distribution of these toxic substances (such as the discovery of DDT in 1965 in penguins in the Antarctic) and their concentration in food chains (such as strontium-90 in mothers' milk) served to symbolize powerfully the potential dangers of global environmental pollution. Perceptions of a radioactive and chemical universe – invisible, all pervasive, insidious – were profoundly disturbing. They not only blighted faith in scientific and technological progress but also compromised the former escape route of geographical mobility, whereby the upper and middle classes had distanced themselves from the older and more local environmental problems arising from slums, smoke and sewage.

People have been distanced from the causes, but not the consequences, of pollution and technological hazards as a result of marked changes in the employment structure. Since the 1960s there has been a decisive shift to the service sector, which now provides most jobs. Consequently, a growing majority of people no longer owe their livelihoods to forms of employment that involve the manipulation or processing of natural resources – whether in agriculture, mining, construction or manufacturing – and that inevitably poses the most acute environmental problems. Fewer and fewer people, therefore, are exposed directly – through their work and income dependency – to a personal conflict between economic and environmental welfare. Opinion surveys do confirm that environmental concern tends to be most strongly expressed by those employed in the burgeoning service sector (Lowe and Rüdig 1986).

Such changes have fostered understandings of the phenomena of pollution separate from its original context, involving a shift in perspective from one to do with the unwanted by-products of production to one that focuses on the consequences of pollution. The more that pollution could thereby be detached from job or wealth creation and could be seen to be deleterious to innocent (and often distant) others (such as residents, children, future generations, wildlife, etc.), the more it could be portrayed as a matter of wrongdoing, that is to say, a moral issue. This has been a specific achievement of the contemporary environmental movement, which has elevated pollution and industrial risks to the status of a crime and forged a new environmental morality (Grove-White 1993).

The development of this environmental morality within Britain has focused particularly on the countryside. This is, in part, the legacy of the rural preservation movement. Since the late nineteenth century the countryside has been seen as a source of spiritual, aesthetic and moral re-invigoration, a view that has propelled a major counter-urbanization movement of middle-class residents throughout the twentieth century, but more particularly since the 1950s. The growing evidence of the ubiquity of pollution has encouraged an even greater sense of protectiveness towards a rural environment, which for many people is experienced as a place of retreat or retirement, detached from productive activities.

Most disturbing of all, therefore, has been a sense of a threat to the rural environment from within. Since the Second World War, agriculture has undergone a profound technological revolution based on an industrial model of high input and high output farming. No other problem so encapsulates the move from simple husbandry to industrialized methods

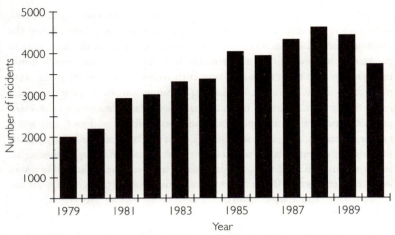

Figure 1.1 Number of farm pollution incidents, 1979–90.
(Source: National Rivers Authority, 1992a: 22)

of farming as does agricultural pollution. In England and Wales, the number of reported pollution incidents from farms more than doubled during the 1980s from approximately 1500 in 1979 to over 3500 a year by the late 1980s (Fig. 1.1), but even the latter figure may be just the tip of the iceberg. From detailed survey work on over 10 000 farms, the National Rivers Authority (NRA) has estimated that "in many catchments the proportion of farms polluting, or at high risk of doing so, is about 40 per cent" (NRA 1992a: 22). Because farm pollution is concentrated in rural rivers and streams, which normally are of a high water quality, and because farm wastes have a strong polluting potential, such pollutants can be particularly damaging. Indeed, in 1990 over a third of all *major* water pollution incidents were from farms (ibid.), leading to the identification of agriculture as "the most damaging single activity in relation to water quality" (Howarth 1992: 53).

A general definition of pollution is that it is matter out of place such that it causes harm or offence. Traditionally, pollution has been associated with industry, and the recognition of agricultural pollution comes as something of a shock. Agricultural pollution is thus both a physical problem and at the same time a dislocation of our conceptual categories, particularly those that place agriculture in a reciprocal relationship with nature and the two of these in opposition to industry.

Approximately 80 per cent of farm pollution incidents during the 1980s were associated with livestock farms, with intensive dairy farms being the most implicated. One farmer and prominent official from the National Farmers' Union even told *Farmers Weekly* magazine "I very much doubt there is a single dairy farmer who isn't causing pollution in some way" (quoted in Paice 1991: 3). The main pollutants from dairy farming are cattle slurry (a mixture of faeces, urine and water), "dirty water" (which contains washings from the farmyard and milking parlour) and silage effluent (fermented grass used as cattle feed). These can be highly polluting when they enter watercourses. Dirty water is, on average, around six times more polluting than raw domestic sewage, cattle slurry 80 times, and silage effluent 170 times more polluting. The threat is vividly encapsulated in a report by the former South West Water Authority, which explained that:

> An average sized Devon dairy farm, with a herd of 53 cows, has a pollution potential load equivalent to that of a community of 465 inhabitants. If silage is made for winter storage, which is likely, then an average crop for a herd of this size would be 650 tonnes. If this crop had been wilted (i.e. cut and left in the field to wilt) then there would be 145 000 litres of silage effluent to be disposed of at the rate of 19 000 litres per day. The potential pollution load of this effluent is equivalent to that of a community of 10 800 inhabitants. (South West Water Authority 1986: 52)

All too often, the diagnosis of environmental problems such as farm pollution is presented as purely and simply a matter of objective science. However, in many environmental conflicts there is confusion and dispute, even among the scientists, about the nature, cause and extent of the problem. What counts as a pollution incident or a quality standard is essentially a social judgement, albeit one that is informed by available scientific evidence. To understand how such judgements arise from contested constructions calls for sociological analysis. Such an approach is particularly appropriate in elucidating agricultural pollution because of the complex social relations and cultural symbols that surround farming and the rural environment. A study of agricultural pollution is, therefore, also a study of the social construction of the environment, and of how different groups struggle to define the specific nature of the problem and construct solutions to it.

It would be wrong, however, to imply that agricultural pollution arises solely with modern intensive production methods and is entirely novel. On the contrary, it would be hard to conceive of a farming system that caused no environmental impact (Conway & Pretty 1991). Very particular pollution problems have long been recognized – for example, the disposal of animal wastes from town dairies. By the mid-Victorian period, too, scientists had recognized that rivers flowing through highly cultivated districts were of poorer water quality and that agriculture was a major source of organic and other impurities, including nitrates leached from manured land (Wilmot 1993). But such problems paled in comparison with sewage pollution and industrial effluents. In any case, it is likely that they diminished during the long agricultural depression that stretched from the 1880s to the 1930s.

The period since then has undoubtedly seen a major and extensive build-up of agricultural pollution. Until recently, this has gone largely unacknowledged as a public problem. Popular perceptions as well as regulatory institutions have focused upon pollution as an urban–industrial phenomenon. When recognition came, it came suddenly, reflecting the charged reaction to any defilement of the rural environment. Legislation quickly followed and in the late 1980s agriculture passed from being one of the least to one of the most formally regulated sectors from the point of view of pollution control.

Such a reversal is a pointer to more deep-seated social and cultural developments, involving, for example, broader reassessments of the social functions of the countryside and the role of farming within it, linked to changes in the social and political structures of rural areas (Marsden et al. 1990, 1993). Above all, it represents the ascendancy of an environmental morality over another older discourse, which lauds farming as a vital and morally worthwhile way of life.

Formal recognition of pollution problems and legislation to address them may, however, be poor indications of what is actually happening "out in society", or "on the ground" (Hawkins 1984). Farming, comprising a myriad of small family businesses, is a particularly intractable sector on which to impose regulations. Equally, established approaches to pollution regulation, oriented towards gross pollution from point industrial sources (i.e. factory chimneys and waste pipes), are not well suited to the regulation of farm effluents (Lowe et al. 1992, Clark et al. 1994).

It is the aim of this book to elucidate these issues, based on a major empirical investigation of water pollution by farming. This investigation was carried out between 1989 and 1995, in the wake of the switch in regulatory law governing farm pollution. At one level, therefore, it is a study of a change in regulatory regime and what that actually entails for regulatory and farming practice. What makes the study of particular interest from this perspective is that the British approach to environmental regulation is often characterized as one marked by informality and an avoidance of legal sanctions; by considerable discretion accorded to those charged with regulatory responsibilities; and by close consultation and negotiation with the industrial interests and concerns being regulated (Vogel 1986, Lowe & Flynn 1989). Within this tradition, agriculture is usually portrayed as representing an extreme case, one best described as self-regulation. Thus, in the past, agriculture has often been given special treatment or exemptions within planning, environmental and safety legislation. Instead, a form of agricultural regulation has long been used which draws upon traditions of autonomy and self-management within the farming community (Cox et al. 1990). The approach relied on the provision of information and advice by farm advisers, the promulgation of so-called codes of practice (i.e. guidelines drawn up in consultation with representative groups specifying good or desirable practice) and specific financial incentives. The central principle was that farmers should be encouraged to adopt a particular kind of conduct, rather than coerced. This voluntary form of regulation is aimed at achieving a practical and farmer-determined balance between the protection of the environment and the commercial production of agricultural produce. The recent shift to a much more formal and legalistic approach to pollution regulation in this, of all sectors, is clearly a momentous event that at one and the same time, raises major questions about how it is being put into practice and portends far-reaching changes in Britain's approach to social regulation.

In practice, what this shift has meant is the insertion into the agricultural community at the local level of pollution regulators armed with new legal powers and under public and political pressure to take a tough stance over farm pollution. Farmers and their advisers have had to adjust to this new presence, as well as to the greater scrutiny and public criticism surrounding their work. At another level, therefore, our study is one of boundary maintenance and change, revealing how a new environmental morality is impacting on a particular occupational community, which for a long while had resisted it. But the change is linked to wider shifts in social meanings and, at its broadest, the study is one of the eclipse of farming as a source of natural values by the new environmental morality.

Tackling these issues has demanded empirical work guided by a methodology sensitive to the differences in meanings and strategies of the various groups involved. The work presented here is a sociological analysis based on a major investigation of the water pollution problems associated with dairy farming. A detailed farm survey has been conducted in river catchments in a locality where pollution problems from dairy farming are rife – Devon, in the South West of England. Interviews and participant observation work have been carried out with pollution control officials and farm advisers in the area. Representatives of farming and environmental interests and agricultural supply companies have been interviewed, along with those responsible for policy-making and for scientific research on agricultural pollution. The book is structured around the ways in which these different actors perceive and evaluate pollution and contribute to the overall social definition of the problem and the formulation of responses and possible solutions to it. The next part of this chapter describes the methodology pursued.

Methodology: following actors, following pollution

Our choice of methodology for researching and analyzing the farm pollution issue has its origins in two main concerns: that the phenomenon of farm pollution and the construction of solutions to it are the product of more than just the farmer's actions; and that the different actors[2] involved in farm pollution may see the problem differently. With respect to the first of these concerns, what dairy farmers do in their fields, cowsheds and milking parlours is to produce goods for the market, and at one and the same time this activity produces effluents, but not necessarily pollution. Weather and topography, for example, intervene such that the same action can result in effluents contaminating a river or stream in one instance but not in another. For contamination to be termed "pollution" it has to be detected, and for this to happen other actors must be involved, such as a vigilant public, the scientists who developed the detection methods and pollution inspectors armed with instruments and knowledge. Repercussions for the farmer are linked not only to the detection of pollution but also to assessments of its seriousness, and interpretations of regulations. This extends the network of actors involved to include regulators, civil servants, politicians and so forth. In addressing pollution risks, farmers' actions may draw in actors as diverse as agricultural advisers explaining control technologies, civil servants deciding on grant applications and journalists reporting on prosecutions in the local paper (Clark & Lowe 1992). Thus, although the farm is a compulsory location in the water pollution story, the processes that lead to pollution are entwined by a whole host of actions, connecting the farmer to myriad other actors near and far, in fields, offices, rivers and laboratories. The methodology we chose had to be able to encompass this assemblage of actors and the interactions that take place between them.

Our wish not to favour any one type of actor but to deal even-handedly with all perspectives was linked to our second concern: how to cope with actors "seeing things differently". A look at virtually any environmental issue will typically turn up a clamour of certainties, opinions, strategies and prescriptions concerning what the real problems are, how they might be rectified and what needs to be done to achieve this. Farm pollution is no different. We wanted to be able to approach our research in a way that allowed us to account for differences in the ways in which actors "see" things, without having to resort to the uncritical elevation of some actors' views or the equally uncritical denial of others. Our interest centred on finding out how actors' worlds are constituted and the actions they provoke; how pollution events, and their solutions, are assembled.

Many sociologists look to society itself, expressed in the form of concepts such as class, institutions, norms, interests, hegemony, and so on, to explain what is going on in the social world. They work with what Bruno Latour (1986) terms an *ostensive* definition of society. Latour contrasts the ostensive definition, in which society itself is the cause of actors' views and behaviour, with what he terms the *performative* definition, in which society is the consequence of what actors do, and in which it is actors themselves who in practice state what society is about.

The performative definition of society thus differs radically from the ostensive definition. It shifts the focus to how society is *made*, in contrast with the ostensive definition's concern with what society *is* in essence. Under the performative definition, the conventional idea of society is turned upside down in the sense that the social concepts (such as social class) conceived as *causes* of social action within an ostensive definition, become *outcomes* within a performative definition. In practice, the performative definition changes the task of the social scientist from that of invoking abstract social concepts to explain why actors do what they do, to that of investigating how actors are linked together in society – how social groupings are forged, built and maintained through time. It does not deny that society exists, but what it emphasizes is how society is constituted.

The two definitions have different implications for the way in which the empirical observation that actors may "see things differently" should be treated. The ostensive definition requires the observer to judge one representation "authentic" and others "mistaken". The performative definition, on the other hand, is concerned about how society's attributes are settled in practice. It can accommodate the reality of "seeing things differently" in a way that does not deny particular rationalities and does not favour particular actors. It is the perspective we have adopted in our study in an approach derived from "actor network theory" (see, for example, Callon 1986a,b, Latour 1987, Law 1992, Murdoch 1994).

Actor network theory espouses a simple analytical principle – "follow the actors" (Callon et al. 1985: 4) – and at times during our study we literally did just that. The task of the analyst is to "describe, with neither fear nor favour, what it is the actors do" (ibid.: 5). This means studying the worlds built by actors on their own terms. Actors construct their worlds from what is around them, that is by designating and associating entities which they select, define and link together. In attributing characteristics to entities in the worlds they build, actors are attempting to speak for others and to impose particular definitions and roles on them. To be successful, other actors' worlds must be colonized. Some actors will be in a better position to accomplish this than others, owing to their control of resources, both cultural and economic. However, success also depends on what other actors do.

At any one point in time, many representations (claims, or, in the terminology of actor network theory, "translations") will already have been successful in the sense that they will have been taken on board by other actors and be embodied in things that circulate between them, such as machines, money, knowledge, regulations and agricultural products (Callon 1991). Other representations will be incipient and uncertain as the actors struggle to achieve them.

The vocabulary of actor network theory makes it possible to explain how actors "define their respective identities, their mutual margins of manoeuvre, and the range of choices which are open to them" (Callon 1986a: 201). Combined with obedience to the principles of "agnosticism" and "generalised symmetry" (ibid.: 200), it allows the analyst to do this even-handedly. No one actor need be favoured, no one perspective taken as superior. The main results of our empirical fieldwork are presented in

Chapters 5–7, and these provide what we hope are quite faithful accounts of the worlds of the actors we studied.

Methods

Our field research was carried out in Devon, and included a farm survey, which was confined to just three river catchments centred upon the district of East Devon. In principle we could have extended our investigations to one or more additional dairy farming areas but this would have meant sacrificing some of the depth and richness of the material gathered. Another possible strategy would have been to undertake a "representative" farm survey (regardless of farm location) while still confining our investigation of other actors to a single area. However, this approach would have lost the advantage of consistent conditions; that is, of all the field actors ostensibly encountering the same circumstances. We thus preferred a fully intensive strategy to a partially or wholly extensive one. However, our choice has also meant that on occasion we have had to address a particular type of challenge: "How do you know that the area you have chosen is typical of dairy farming's pollution problems?"

The question implies that, if the area is not typical, then the findings of the study cannot be extrapolated from the particular setting to the general issue of farm pollution. This sort of challenge to case studies has been addressed by Mitchell (1983). He argues that questions about typicality betray, in the mind of the questioner, "a confusion between the procedures appropriate to making inferences from statistical data and those appropriate to the study of an idiosyncratic combination of elements or events which constitute a "case" (ibid.: 188). This confusion derives from failing to distinguish surface relationships (correlations) and logical connections between features of a situation. While the validity of extrapolating correlations from a sample to the whole population – for example, asserting that large farms are in general more profitable – does depend on the representativeness of the sample, the validity of inferences concerning the processes that link such features does not, but depends instead on the cogency of the reasoning. For example, invoking economies of scale to explain a link between profitability and farm size would probably be accepted; invoking farmers' religious beliefs would probably not be. However, in neither case is the reasoning related to the representativeness of the sample; describing a relationship does not involve the same sorts of procedures as explaining it.

Questions about typicality, however, betray more than confusion over analytical practice. They also rest on the basic assumption that the pursuit of "the typical" is a meaningful quest. In the case of farm pollution, this assumption is highly dubious. Our initial work quickly revealed the heterogeneous constitution of pollution problems. Pollution is locally composed and locally specific, reflecting the heterogeneity of the natural world in which it occurs and the diversity of the social contexts through which it is represented. There is no such thing as an area or farm that typifies pollution, and hence our decision to focus on a particular geographical area, and our choice of area, reflects rather different criteria.

We selected a part of the South West as the location for our study because this region appears *archetypal* of the farm pollution issue. This is not just a matter of numbers, although recorded farm pollution incidents do in fact put the South West at the top of the league. More importantly, the South West served as the prototype for the expression and construction of farm pollution issues in a national context. By the mid-1980s such pollution, involving dairy farms in particular, had become a clear issue there, preceding and foreshadowing similar concerns in other regions.

Our choice of Devon was determined primarily by the size and significance of the dairy industry there. It is the most important type of farming in terms of acreage, number of farms, farm profitability and most other indicators. In a national context, Devon's dairy farms tend to be smaller than average; at the time of our study the average herd size was 60 cows, while the figure for England and Wales as a whole was 65 cows. Small producers present perhaps the greatest challenge to most types of agricultural regulation. The county also has a relatively wet climate and varied topography, whose most striking features are the range of heights encountered, the frequency of steep-sided valleys and the abundance of watercourses. The archetypal Devon dairy farm is sited on the side of a hill with a stream running beside, below or even through the farmstead. Few settings are worse in terms of the pollution threats they pose.

Although the Devon environment is particularly vulnerable to farm pollution, the characteristics of that environment — its mild climate, its varied topography, its small-farm landscapes — have proved very attractive to outsiders. Since the 1970s, Devon has seen some of the highest levels of population inmigration of any county. Rural Devon, in particular, has experienced a large influx of commuter families, retired people and second homeowners. Such people often have a different perspective from that of farmers, not only on issues to do with agriculture and the environment but also on the appropriateness of regulation or the law to deal with them. It was felt therefore that the area presented a potentially fruitful one in which to examine how the social construction of pollution and the regulatory response was shaped by social change.

Following our decision to work intensively within a limited geographical area, we embarked upon a series of linked studies that investigated the intersecting "worlds" – the views, perceptions and attitudes – of key actors in the field: farmers, Pollution Inspectors and agricultural advisers. This called for close and empathetic observation of the encounters between them and the discourses through which these encounters were constructed. We therefore employed participant observation, shadowing the field staff of the National Rivers Authority (NRA) and the Agricultural Development and Advisory Service (ADAS) in their dealings with farmers on pollution matters. To understand the responses of the farmers more fully, we also conducted an in-depth survey of 60 farms in the area. Finally, semi-structured interviews with policy actors, including regional and national officials and pressure-group members, and analysis of documentary sources, were used to explore how developments in rural Devon had influenced, and been influenced by, national developments in policy and politics.

The research was carried out at a time which all the actors involved recognized as being one of great uncertainty. In recent years, agricultural policies and priorities have been changing rapidly, and even now stability is not in sight. At the same time, the dominant public image of farmers has changed from that of the "guardians of the countryside" to "destroyers of the environment". Indeed, farmers see themselves now as the whipping boys of a vocal public increasingly demanding that the rural environment be made over in their own idealized images. Public and political pressures to clean up the environment have been accompanied by increasingly stringent formal and legalistic controls over farm effluents; regulatory arrangements that contrast markedly with past practice characterized by rather more relaxed, voluntary and informal arrangements. When we began our project, the NRA was but a few months old; ADAS was continuing in the throes of radical structural and functional change that were making it more subject to commercial pressures and accountable policy targets; and the regulations to control livestock and silage effluents were going through parliament and their implementation began during the research period.

For the observer, attempting a definitive account of such fluid and determinate circumstances could be the stuff of nightmares. From our perspective, where interest centres on how phenomena and events are constituted, a study at this point in time offered an ideal opportunity. It is at such times that boundaries change, meanings shift, and social and institutional relationships are restructured. Watching the story as it unfolds can reveal the formative processes, the struggles, the choices made, which subsequently become submerged when circumstances have become more stable and all the trials and tribulations of the time have hardened into opaque truths and normalities.

Outline of the book

The study focuses on how the phenomenon of pollution and ways of solving it were constituted. How did the various actors define pollution, and how did they see the farm pollution problem? How did each see the other actors involved? Who was managing to enrol whom into their way of "seeing the world"? Whose version was succeeding?

We begin by describing the context in which farm pollution became a public issue. Chapter 2 outlines how dairy farming in Britain has changed over recent decades and in doing so has created certain pollution risks. In Chapter 3 the traditional official response to pollution from farm effluents is described. Informed by the view that this was an unfortunate but infrequent concomitant of a progressive agriculture, the response was to treat pollution as a minor technical matter to be left to the industry, and as being of no wider public concern. Over the years, a variety of codes of practice to protect the environment were promulgated, covering a range of farming activities, and these were backed up with information and advice on pollution control from ADAS, the farming advisory arm of the Ministry of Agriculture, Fisheries and Food. In addition, grant aid could be obtained for facilities to store and handle farm effluents installed under farm development schemes. At the same time, however, the much more powerful stimulus of production policy acted to increase pollution risks.

The situation altered dramatically in the mid- to late 1980s. Evidence of mounting farm pollution incidents and of deteriorating river quality helped politicize the problem of farm wastes in a political context charged by the prospect of water privatization and by the challenges to the narrowly productivist aims of agriculture following the European Community's imposition of milk quotas.[3] These two developments opened up respectively the fields of water pollution and agricultural policy to external scrutiny and public debate. Water pollution from farm effluents, lying at the intersection of these two previously contained policy fields, was subjected to an unprecedented glare of attention. No longer was it possible to maintain that this was simply a technical matter. Other definitions of the problem were able to emerge that stressed the primacy of environmental protection rather than production and which, at their most radical, presented the issue of pollution from farm effluents as an indictment of an over-intensive agriculture. In 1987, a parliamentary Select Committee concluded that rising pollution from farm effluents in Britain was an important contributory factor to declining river quality, and called for "a far more interventionist and regulatory approach to farm pollution" (House of Commons Environment Committee 1987). Chapter 4 outlines the circumstances of this politicization of the farm effluent problem and the resultant shift in the regulatory regime surrounding it.

The Pollution Inspectors of the NRA were thereby charged with new powers to tackle pollution from farms. Chapter 5 examines how they actually exercised this authority in the field. The Pollution Inspectors' work and outlook are portrayed, and their attitudes towards agriculture and water pollution are described, providing a basis for considering the strategies they adopted in relation to farmers.

Pollution control was one of many pressures faced by the farmers. Although they acknowledged problems with farm wastes, they resented being stigmatized as polluters. To them, practical and financial constraints loomed large in considering how to respond to the pressures upon them. Many sought to draw the Pollution Inspectors into an understanding of these constraints. Chapter 6 therefore examines the farmers' perspectives on farm pollution.

Other groups of actors were also involved in negotiating pollution control on farms, including agricultural advisers, environmental activists and concerned local people who were prepared to challenge what they considered to be unacceptable practices. Chapter 7 examines the emerging social networks around local pollution control as part of the changing rural world.

Our strategy in writing this book has been to "describe, then analyze." The final chapter draws together our analysis from our field research. It concludes by discussing how the problem of pollution from dairy farms challenged the conventional representations of pollution and agriculture, and how its regulation pitched together new and old conceptions about responsible conduct towards the natural environment.

Notes

1. Here the term social regulation is used as defined by Hawkins as "those forms of regulatory control that are not directly concerned with the control of markets or other specific aspects of economic life, but instead aim to protect people or the environment from the damaging consequences of industrialisation" (Hawkins 1989: 663).
2. We use the word actor in a generic sense, meaning any entity which can represent other entities and act in society (Callon 1986a). However, the essence of our usage is that actors' worlds and actions are collectively generated even though it is ultimately individuals who act.
3. Agricultural productivism is taken here to imply "a commitment to an intensive, industrially driven and expansionist agriculture with state support based primarily on output and increased productivity" (Lowe et al. 1993: 221).

CHAPTER TWO

Changing dairy farming and the pollution problem

The practice of dairy farming has changed radically over the past 40 years or so. Technological change and the economics of agricultural support have encouraged the concentration of production, which in turn has driven increases in farm size. Together these interlinked developments have created pollution risks on the modern dairy farm. Farmers now have to dispose of large volumes of liquid effluents of great potency where once they returned solid manure to the land as a crop nutrient. The developments in the scale and techniques of dairy farming that have brought about this situation are described in the first half of this chapter. Whether or not pollution actually occurs in any particular instance and, if so, the degree and extent of the consequent damage, depend upon more immediate factors. So, the chapter goes on to consider the routes by which farm effluents may come to contaminate rivers or streams and the implications for the water environment.

Processes of change on dairy farms

Specialization, concentration and changing farm size
In order to achieve economies of scale, farmers have tended to concentrate their land, labour and capital resources on the production of a smaller range of commodities. Given the high relative capital costs of milk production, for many dairy farmers this has meant a move away from mixed livestock and cropping systems and an emphasis on expanding the dairy enterprise. As a result, the average size of dairy herds in Britain has more than trebled since 1960, when it was just 21 cows.

More fodder has been required to feed the extra cattle, so either more land has been needed or existing land has had to be farmed more intensively. Until the 1980s the strategy of investing surplus profit into buying extra land was an important element in a farming ethos, fuelled by rising land prices. At the same time, the rising real cost of farmland encouraged farmers to increase the density at which their farms were stocked.

The switch from hay to silage as the main feedstuff, combined with improved fertilizer techniques and grass seed varieties, did mean that farmers were able to increase the number of cows kept on the same area of land. Thus, the rate of increase in *farm* size did not keep pace with that of *herd* size. According to Milk Marketing Board statistics, the average land area of dairy farms in England and Wales increased by 75 per cent between 1960 and 1987, but the amount of land per cow decreased by 43 per cent over this period. In consequence, there was more farm effluent to be stored and less land to spread it on. Slurry-spreading has gradually become more a practice of "waste disposal" than simply the manuring of grassland, an outlook reinforced by the availability of cheaply priced and easily usable manufactured nitrogen.

For most of the post-war period, such growth and specialization were officially encouraged. However, this changed in March 1984 with the European Community's introduction of milk quotas, targeted to curb milk supply and tackle the budgetary crisis of the Common Agricultural Policy. The Ministry of Agriculture, Fisheries and Food (MAFF) referred to the move as "a seismic shock", a view shared by dairy farmers who found their milk production was to be restricted, almost overnight, to 9 per cent less than their 1983 production level. It was a policy change that was especially hard to swallow for those producers who, in line with government policy, advice and grant incentives, had been modernizing and expanding, often with money borrowed against assumed increases in output.

Specialization of production gives the farmer less scope to adapt in response to changing external conditions. Those farmers who had concentrated resources into the dairy enterprise and expanded its scale prior to the 1980s often found themselves locked into this strategy. Moreover, the course of action now favoured in official circles – of diversification into non-farming work – is less feasible an option for dairy farm households, who tend to be more dependent on agriculture for their income than other farm households, not least because of the particular daily demands that dairy farming makes on family labour (see Ch. 6).[1]

In consequence, the imposition of milk quotas has had little impact on the tendency towards specialization and concentration in the dairy sector.[2] What has happened is that increasing the intensity of production and self-sufficiency in feedstuffs has become more important than expanding farm size or herd size to achieve economies of scale. Obtaining additional milk quota is more important to farmers than buying more land or stock, particularly with cuts in quota and continuing increases in milk yields per cow.[3] Data from MAFF's annual June census show how stocking rates of dairy cows changed during the 1980s. The average increased by 10 per cent in Devon, an increase slightly greater than that for England as a whole.

Enlarged herd sizes, along with the growth in stocking densities, have meant not only that more cattle require feeding, bedding and watering but also that greater volumes of effluent have to be dealt with. The greater quantity of slurry has implications not only for pollution risk from the farmyard and buildings but also in the fields. In considering whether any farms simply have too much livestock for their land area, it should be borne in mind that the vast majority of dairy farms carry not only cows but dairy followers (the young cows that will eventually become milkers) and often other cattle as well. Using the figures in MAFF's *Code of practice for the protection of water*, specifying the land area required to spread effluents from different types of livestock safely, overall stocking rates have been derived for the 60 dairy farms we surveyed in Devon (MAFF/WOAD 1991: 69). In 1981 43 per cent of the farms we surveyed were stocked at rates higher than three dairy cow equivalents per hectare and by 1991 this had risen to 53 per cent.

The crucial constraint, though, is not the total size of a farm, but the area within it that is suitable for spreading effluents. The Ministry's code of practice, for example, states that suitable land should be at least 10 m away from any watercourses and 50 m from a spring, well or borehole, and that steep sloping land should be avoided (MAFF/WOAD 1991: 7), which significantly reduces the area suitable for spreading on most Devon farms. Current MAFF guidance is that the minimum area of land required for disposal of the effluent of one cow is 0.16 ha (ibid.). On this basis, it is likely that most of the 18 per cent of dairy farms in our survey with overall stocking levels in excess of four cows per hectare are overstocked. The NRA certainly believes that livestock farmers have intensified their production and increased herd sizes "to a point where the associated land is often insufficient to cope with the slurry produced" (NRA 1992a: 33). The EC Nitrates Directive 1991, which is concerned with protection of

Table 2.1   Production levels of animal manure by dairy farms
in the West of England 1990/1991

	Average Production of Animal Manure (kgN/ha)	Share of Total Number of Farms (%)
Farms with nitrogen from manure not exceeding 170 kg per hectare	127	58
Farms with nitrogen from manure exceeding 170 kg per hectare	211	42
Total number of dairy farms represented (11 120)	156	100

Source: Farm Accounts Data Network (FADN) European Commission: Analysis by the Agricultural Economics Research Institute, the Netherlands (LEI-DLO).

waters against nitrate pollution from agricultural sources, requires that the application of animal manure in vulnerable zones should not exceed 170 kg of nitrogen per hectare. Table 2.1 presents estimates of the proportion of dairy farms in the West of England (including the South West) that exceed this level of production of animal manure. Of course, not all these farms will be in areas vulnerable to water pollution.

Technological changes

Technological changes in dairy farming have been closely related to processes of specialization and the concentration of production. The technological developments that have been instrumental in raising pollution risks are the move away from housing cattle on straw in loose housing to slurry-based systems with concrete-floored cubicles and the trend towards making silage rather than hay. With these innovations have come associated changes such as the concreting of farmyards. At the same time, the use of water has increased markedly. Indeed, the dairy sector is the largest agricultural user of water (Bailey & Minhinick 1989).

Traditional loose housing of cattle on straw beds yields solid farmyard manure (Fig. 2.1). In contrast, many slurry systems use little bedding, and so the waste washings from these consist largely of a mixture of urine and faeces, plus spilt food and water with only a small amount of straw or other bedding matter. Overall, the solid material usually makes up less than 15 per cent (Mason 1992). Slurry systems were first introduced in

Figure 2.1 Traditional solid farm manure is easy to stack in heaps and poses less of a pollution risk than liquid effluents. (Photograph courtesy of Louise Morriss)

Figure 2.2 Dairy cows housed in concrete cubicles with an automatic scraper to remove slurry. (Photograph courtesy of the Water Services Association)

the early 1960s. Dairy farmers find that they not only allow more cows to be kept in the same shed area but also that they reduce the workload for mucking out the housing. This is a chore that many farmers and labourers find tedious and which, moreover, is also perceived as a poor use of time. Slurry's liquid form makes it more amenable to mechanical removal from buildings and yards, and automatic scrapers (Fig. 2.2), pumps and tractors can be used. With some systems the slurry is simply scraped or hosed out of the housing. Others have slatted floors over underground reception pits or transfer channels, from which the slurry can be sluiced, scraped or pumped. The Royal Commission on Environmental Pollution estimated that in 1976 just over half the dairy farms in England and Wales used some type of slurry-based system (Royal Commission on Environmental Pollution 1979: 129). By 1989 the proportion had risen to two-thirds, and in our survey of Devon dairy farms in 1991 it was 82 per cent of farms (but, because these tended to be the larger farms, they accounted for 88 per cent of the cows covered by the survey). With the typical dairy cow producing around 57 litres of slurry per day (MAFF/WOAD 1991: 69), the average dairy herd in our sample (78 cows) would produce about 800 000 litres of slurry during the six months of the year for which herds are normally kept inside.

Modern housing systems and milking parlours also use greater volumes of water for cleaning. In part this is because the regular washing down of milking parlours is stipulated in the Dairy Regulations, which set statutory hygiene standards and which have become more stringent over the years. At the same time, the greater convenience of high-pressure hoses has made them a common means of washing down equipment and hosing out cubicles and yards. For the average herd in our sample, such use of a power hose would mean over 2700 litres of washings to be handled each day. These parlour and yard washings include faeces, urine, waste milk and the chemicals used for sterilizing parlours. They are commonly referred to as "dirty water", a term that also includes rain runoff from dirty areas of the yard. The pollution potential of dirty water is lower than that of slurry but high enough to have a serious impact on watercourses (see Table 2.2 below).

Table 2.2 Biochemical oxygen demand (BOD) of common organic wastes

Effluent type	BOD range (mg/l)[a]	Typical BOD (mg/l)[b]
Treated sewage effluent	20–60	
Raw domestic sewage	300–400	350
Dirty water[c]	1000–2000	2000
Cattle slurry	10 000–20 000 ⎫	
Pig slurry	20 000–30 000 ⎬	30 000[d]
Poultry slurry	30 000–35 000 ⎭	
Silage effluent	30 000–80 000	60 000
Milk	140 000	140 000
Clean river water		<5

Notes
a. *Source*: MAFF/WOAD (1991: 3)   b. *Source*: NRA (1992a: 25)
c. Dairy, parlour and yard washings   d. Figure for "animal slurry".

The concreting of yards has accompanied the move from farmyard manure to slurry systems, and increases in herd size have often meant enlargement of the buildings in which the cows are housed and fed. The consequence is a greater area of impermeable surface, which in turn increases the volume of rainwater that can collect through runoff. In the wetter parts of Britain, there will be around a thousand litres of runoff for every square metre of concrete per year (Paice 1991). Yards and waste storage facilities are not normally covered, guttering is commonly inadequate, and separate drainage for clean and dirty water is often lacking. Therefore, many farmers are unable to exclude rainwater runoff from their polluting effluents, which greatly increases the volume of contaminated liquid to be managed and effectively diminishes the capacity of effluent storage facilities. This in turn means that periods of heavy rainfall put these facilities under the greatest pressure at the very time when farmers should be storing effluent rather than disposing of it on land. At any time of the year, though, a sudden downpour may overwhelm inadequate storage systems.

Various types of storage facilities are used for farm effluents. The smallest ones are simple pits; for most farms these are too small, although for some they represent the only storage available. Other types generally hold much larger volumes. Lagoons (shallow holes with earthed banks) and large cylindrical "tin tanks" (or "above-ground stores") are used to contain thin slurries. In the tanks, which are commonly regarded as providing the most secure containment (Mason 1992), the slurry is normally stirred to prevent separation of the solid and liquid fractions, and pumped into a slurry tanker for disposal (Fig. 2.3). In contrast, a common practice with lagoons is to use a strainer box to separate out the solids, facilitating the disposal of the liquid by low-rate irrigation or tanker. "Weeping wall" stores are used for thicker slurries; typically they have concrete floors and sides constructed of railway sleepers or concrete panels; narrow gaps are left in at least one side through which liquids, but not solids, can pass to a separate effluent tank (Fig. 2.4). Slurry separators are also available, which mechanically separate the solid and liquid fractions. The solid waste can then be handled using a manure spreader, and the liquid can be disposed to land by low-rate irrigation.

Dirty water can be handled with the slurry. Where handled separately, it is typically piped to settlement tanks. The small proportion of solids settle out as sludge and the liquids are piped to a field irrigator, which is normally run continuously at a low rate during the daylight hours.

Figure 2.3 An above-ground store (or "tin tank") with slurry being emptied into a tractor-drawn tanker to be spread on farmland. (Photograph courtesy of Graham Cox)

Figure 2.4   A newly constructed "weeping-wall" slurry store.
(Photograph courtesy of Graham Cox)

Using grass and crop land for effluent disposal and treatment is likely
to remain the predominant method, in part because manure, slurry and
even dirty water can still serve as nutrients. Certain active treatments have
been tried, including mechanical aeration, anaerobic digestion, reed beds
and barrier ditches. All have their costs and drawbacks; none are capable of
producing an effluent that is anywhere near clean enough to be discharged
into a watercourse, and consequently their use is rare (Mason 1992).

Official advice on the design and construction of slurry and dirty-water
systems has been available since the mid-1960s (Agricultural Research
Council 1976, MAFF 1980, 1982a,b, 1983). The most recent publications
at that time formed part of the recommendations contained in the 1985
Code of good agricultural practice (MAFF 1985). However, adherence to any
such guidance was entirely voluntary until the introduction in 1991 of
regulations under the 1989 Water Act (*The control of pollution (silage,
slurry and fuel oil) regulations* 1991). These "Farm Waste Regulations" set
statutory technical standards for the design and construction of *new*
and *substantially altered* slurry and dirty-water handling facilities. Such
facilities must be impermeable, must be guaranteed to last for at least 20
years with proper maintenance, must be sited at least 10 m from any

watercourses (including ditches), and must incorporate at least four months' storage unless the farmer can show that adequate provision can be made for disposal to land. There is also provision for ordering the upgrading of existing installations should a serious pollution risk be apparent. The Water Act also required MAFF to issue a new *Code of good agricultural practice*, which it did in July 1991. Unlike the Regulations, failure to keep to the Code does not constitute an offence, but such a failure may count against the farmer in legal action over water pollution. The Code is intended as "a practical guide to help farmers and growers avoid causing water pollution" (MAFF/WOAD 1991: 1) and covers not only slurry and dirty-water management but also silage effluent, fertilizers, pesticides and other potential pollutants.

However, the majority of systems currently in use were built before the new Code was issued and the Regulations came into force. Previous guidance not only carried no statutory weight but was in any case less thorough. First, lower figures were used for calculating storage capacity than the now accepted 57 litres of effluent per cow per day. Assuming it was properly estimated at the time, storage capacity would be insufficient by 25–40 per cent by today's standards (MAFF 1980, Mason 1992: 26). Secondly, although the guidelines included more construction advice over the years, and alluded to technical standards – for example, that slurry stores should be built in accordance with (the old) British Standard 5502 and that earth-banked structures should be impermeable (MAFF 1982b) – in practice few farmers are thought to have complied. Constructing proper effluent management facilities was simply not perceived as a productive use of available capital. Rather, the arrangements made tended to be determined primarily by the "cow end" of the system, with makeshift facilities put in at minimal cost and least disturbance to daily farming practices. Any design and construction generally relied on no more than the skills of the local jobbing builder or the individual farmer's do-it-yourself capabilities. It is only very recently that pollution prevention concerns have come to the fore in the design and construction of farm waste facilities.

Decades of dealing with livestock effluent as cheaply as possible have left a legacy of inadequate handling systems. Some 14 per cent of British farms handling slurry and 34 per cent of those handling manure have no storage whatsoever (Furness et al. 1991: 28), necessitating spreading on the fields whatever the weather. Moreover, more than three-quarters of all farms have less than the four months' storage capacity recommended by the regulations. Many, if not most, older systems are prone to leakage

Figure 2.5 A dairy cow grazing on silage, with effluent escaping into the ground. (Photograph courtesy of the Water Services Association)

– for example, lagoons and pits built on permeable soil and without liners make seepage unavoidable. Poorly laid concrete, which corrodes relatively quickly, has led to chronic leakage, and weakened structures have a higher risk of catastrophic failure. Dirty water, too, is often overlooked and seldom separately managed: about a quarter of farms have no provision for storing parlour and dairy washings, and almost two-thirds have no store for yard runoff (Furness et al. 1991: 22, 24).

In short, effluent management has not kept pace with the new demands of slurry-based systems. Poor quality handling facilities, insufficient storage and poor spreading practices intensify the likelihood of livestock effluent escaping into watercourses, increasing the risk of pollution. The farm pollution stakes have been raised even further by the second major technological development of the times: the trend towards a greater use of silage rather than hay as the main source of winter fodder (Fig. 2.5).

The advantages for dairy farmers in making silage are twofold: it is less vulnerable to unpredictable weather conditions than is hay-making; and it can also be cut earlier, making several cuts possible in the season. Although the switch at first meant farmers had to invest in a silage clamp or silo and learn new techniques, the benefits in terms of convenience,

Figure 2.6 Big-baled silage, wrapped in polythene, stacked in a farmyard. (Photograph courtesy of Graham Cox)

labour saved and increased feed production led to widespread adoption of the technology (Brassley 1996). Subsequently, the introduction of big balers, which wrap the silage in polythene, enabled many smaller livestock farms to make silage without the need to invest in an expensive silo (Wilkinson 1990) (Fig. 2.6).

Additives are available to assist silage-making. Wet grass is prone to rotting and makes poor-quality silage. This can be avoided by additives that promote proper fermentation and preservation. So, their use renders silage-making less vulnerable to wetter weather. Early products, such as molasses and salts of formic acid, have been joined on the market by a wide range of additives, including sulphuric acid and microbiological innoculants. However, a significant minority of surveyed farmers disapproved of the use of silage additives, feeling that they represented a "short cut" or "easy option". As one farmer explained to us, "If weather conditions are such that a silage additive is required, you shouldn't be making silage."

Silage-making itself produces an effluent that is extremely polluting, up to two hundred times more so than untreated sewage (see Table 2.2 on p. 24). The effluent is also difficult to control; its acidity makes it corrosive, and it can easily escape the confines of the silo. The quantity of silage effluent produced depends on how wet the grass is when it is ensiled, and whether an absorbent such as chopped straw is used. Ideally, silage effluent should be intercepted in collection channels (including channels around the silo perimeter to catch any escaping liquid) and drained to a purpose-built, corrosion-resistant tank, from which it can be spread or irrigated on land. The Farm Waste Regulations specify that new tanks must be impermeable and made to last for 20 years without maintenance. The Regulations also set down a minimum tank capacity, which is approximately equivalent to 30 litres for every tonne of silage made.

These new statutory requirements represent considerably higher technical standards than obtained in the past when the majority of silos now being used were built. A survey carried out in Scotland in 1981 (Brownlie & Taylor 1982) clearly exposed the shortcomings of silo construction at that time. Defective floors were found to be the most common cause of effluent leakage, and fewer than half the silos surveyed had separate effluent tanks. These were often too small for the amount of effluent produced and defective drainage from silos to tanks was also a significant problem. The study concluded that most of these problems arose from design flaws, use of unsuitable materials and faulty workmanship in construction. Information for Britain as a whole is provided by a more recent survey (Furness et al. 1991). Its findings suggest that some 81 per cent of dairy farmers who make silage use a silo. Of these, about half collect effluent in a separate tank, but almost half of these tanks hold less than the statutory minimum capacity. Of those farmers without tanks for silage effluent, more than 20 per cent take no action at all to deal with it, and others fail to take care of all the effluent produced (ibid.: 156). The picture becomes even more grim when the likelihood of poor design and construction is considered. The Scottish survey found that 70 per cent of the 50 randomly selected silos surveyed had structural defects (Brownlie & Taylor 1982). Thus, a substantial proportion of existing silos may be liable to leak, regardless of whether collection and storage facilities are sufficient in practice to cope with the quantity of effluent produced.

Although the state of existing effluent systems on farms gives cause for serious concern, the farmer whose system is substandard will not necessarily precipitate pollution. Conversely, the farmer with a technically sound system will not necessarily avoid polluting. Whether effluent escapes depends not only on the quality of the technology but also on how well the system is managed and maintained. Then, whether a significant amount of effluent reaches a watercourse is related to the location of the farm, the topography of the countryside and the exigencies of the weather. The majority of farmers rely on spreading and irrigation as a means of final disposal, and these practices may also lead to pollutants reaching watercourses. Furthermore, the extent of any pollution so caused depends on the nature of the particular effluent and its rate of discharge. So, the pollution risks posed by a modern dairy farm are complex.

Effluents in the environment

Routes of contamination

The deliberate discharge of farm effluents into watercourses is now quite rare, although it was once a common practice (see Ch. 3). Even so, it is estimated that some five per cent of dairy farms nationally still discharge dairy and parlour washings straight into a field drain or stream, and a similar percentage send dirty water runoff directly into a ditch (Furness et al. 1991: 22, 157).

The routes by which effluents can unintentionally escape are various. One already mentioned is overflow from systems too small to cope with heavy rain. Like all engineered structures, effluent facilities are not contrived to cope with every eventuality. In particular, the storage capacity may cope with rainfall only up to a particular intensity. A so-called return period of, say, five years means that a rainstorm that will overwhelm the system will occur once every five years on average (Mason 1992). The extent to which older installations have been designed to allow for storms is uncertain, but in any case the choice of return period involves a trade-off between the additional cost of more capacious storage and the consequent reduction in the risk of pollution from overflow.

Leaks and accidents can happen at any time. The less charitable might term many of the latter "management failures"; valves are left open, pumps fail to work, silage effluent tanks overflow, transfer pipes come uncoupled, tanks burst. Some systems include engineering controls such as lights or bells to raise the alarm when things go wrong, but these too may fail to operate properly, or all too frequently are deliberately turned off because of irritating false alarms. Chronic seepage because of failure to maintain a system that was originally perfectly adequate might also be termed a management failure, as might seepage attributable to poor design and construction, on the grounds that a good manager would do something about it. On the other hand, such seepage may be hidden from view, and it can percolate through the ground to emerge some distance away. Similarly, imminent catastrophic failure, such as a tank in danger of bursting, may not be obvious to someone who is not trained to recognize a critically weakened structure.

Nonetheless, and although requirements for management and maintenance are bound up with the quality of the system, unintentional escape of effluent is often a reflection of the farmer's management capabilities. The tasks of operation are rarely difficult, but they are routine and onerous in the sense that continual vigilance is required if accidents are to be avoided. At the same time the demands of animal husbandry, silage-making, paperwork and the like compete with technical duties; the needs of the cows and the business may well take priority over, say, checking the pump or getting out with the spreader to avert a potential overflow. Similarly, repairs and routine maintenance may be put off until what the farmer perceives as more pressing matters have been dealt with, although the fact that the tasks involved are unpleasant and may be hazardous may also encourage procrastination. Among ADAS and NRA field staff, farmers are notoriously perceived as indifferent (or even incompetent) managers of equipment. Operational failure, bursts and leaks have caused the vast majority of recorded pollution incidents involving slurry, manure and silage effluent (Table 2.3). In turn, these made up more than 30 per cent of all pollution.

The disposal of effluents to agricultural land provides the second major route by which pollutants can escape inadvertently and, in recent years, incidents caused by land runoff have made up to about one-sixth of recorded incidents. Unlike design and construction, disposal is not controlled under the Farm Waste Regulations. Rather, reliance is placed on farmers adhering to the *Code of good agricultural practice*, which encourages them to plan when and where they will spread effluents. The risk that effluent will reach a watercourse and cause pollution is increased if it is spread during wet or freezing weather, at too high a rate of application, on sloping land, on land adjacent to watercourses, on permeable or cracked soils overlying an aquifer and on fields that are under-drained. It is somewhat ironic that fields with an effective system of land-drains present an increased risk, given that up until the 1980s much subsidy was poured into land drainage in the name of agricultural improvement.

Farmers' spreading practices were surveyed just before the introduction of the current *Code of good agricultural practice* (Furness et al. 1991). It was

Table 2.3 Farm pollution incidents involving dairy cows/cattle by source, 1989

Source of pollution	Reported incidents	
	number	%
Slurry stores		
inadequate storage design capacity	203	10.3
bursting/leaking stores	171	8.6
man-made holes/sluices/valves	40	2.0
poor storage operation	175	8.8
Solids stores		
inadequate storage capacity or containment	68	3.4
poor storage operation	53	2.7
Yard washings	395	20.0
Dairy/parlour washings	183	9.3
Land run-off	380	19.2
Treatment system failure	65	3.3
Silage		
inadequate effluent storage	70	3.5
leaking silos	100	5.1
leaking effluent stores/drains	75	3.8
Total	**1978**	**100.0**

Note: This is the latest year for which such a detailed breakdown of the statistics is given.
Source: National Rivers Authority (1990b: 21).

estimated that, on more than half of all dairy farms, slurry and manure are spread on land adjacent to rivers, brooks or ditches, with 10 per cent doing all their spreading on land next to watercourses. Just under half were estimated to utilize sloping land, with 5 per cent using steeply sloping fields. Some 40 per cent of farmers interviewed also believed that lack of storage capacity forced them to spread in unsuitable conditions (ibid.: 32). However, even where enough storage capacity exists, effluent disposal is often adjusted to the convenience of the farmer, rather than related to the riskiness of the operation in terms of pollution. Of course, the two considerations might fortuitously coincide. The Devon dairy farmers we surveyed were asked what the most important factors were in deciding when to spread slurry. The majority cited ground and weather conditions; but, for most, their logic turned out to be agricultural rather than environmental, explaining their response in terms of a concern not to damage the grass sward with heavy machinery.

Environmental impacts

Before discussing water pollution implications, it is worth noting that, although farm effluents have their greatest impact on the water environment (Conway & Pretty 1991, National Rivers Authority 1992a), they can also damage other habitats (Nature Conservancy Council 1991). Pathogens, too, lurk in animal effluents and it is possible for them to be transmitted to humans via polluted watercourses – for example, the parasite that causes the disease *cryptosporidiosis* may be ingested in contaminated drinking water or during water-based recreation (Milne 1989). In addition, where livestock farms are located near villages or towns, odour from storage facilities and spreading can become a source of complaint (Nielson 1990).

The seriousness of farm pollution depends on the strength of the effluent, which is commonly gauged by its biochemical oxygen demand (or BOD). This is a measure of the amount of oxygen required to break down organic matter in the effluent and, as such, it indicates the potential impact of a discharge into watercourses in terms of oxygen depletion. Table 2.2 provides estimates of the BOD of some organic wastes. It includes sewage effluent to show the much higher pollution potential of livestock as compared with human excreta. The range in BOD value for each type of material reflects the variability encountered in practice; substances of relatively consistent composition, such as milk, show very little variation.

Organic wastes disturb the water environment in three main ways (Hellawell 1986, Haslam 1990, Holmes 1990, Nature Conservancy Council 1991). First, concentrated effluent can kill organisms directly; in sufficient concentrations, ammonia (from excreta) and chlorine (from the hypochlorite used to sterilize milking parlours) can be toxic to both fish and invertebrates. Secondly, the decomposition of organic material reduces the amount of dissolved oxygen in the water, which can lead to death by suffocation of species intolerant of lowered oxygen levels. If oxygen depletion is severe, it can even cause the death of micro-organisms that feed on waste, thus breaking the self-purification process and rendering water completely lifeless. Thirdly, nitrogen and phosphorus in the effluent can change the nutrient balance of the receiving water. Plants that tolerate nutrient-poor conditions are particularly vulnerable, as they can easily be suppressed by more competitive species, which flourish in enriched conditions.

The net result of these processes is a reduction in the diversity of plants, fish and invertebrates, with only species tolerant of the particular level of contamination being able to survive in the water. The precise impact in particular places will depend not only on the type of effluent but also on how much effluent enters the water, how rapidly it gets there and the character of the receiving water. The pattern of discharge may assume any permutation from a slow trickle to a rapid flood, and discharges themselves may be solitary, intermittent or continuous. Correspondingly, impacts range from transient to persistent. With large water bodies, effects can be mitigated by dilution, whereas fast flowing water can carry pollutants down stream. Flowing water also gives mobile organisms an opportunity to escape contamination. Variations in natural water chemistry and the presence of other pollutants – such as domestic sewage, fertilizers from leaching or runoff (which raise nutrient levels) and sulphur and nitrogen compounds from atmospheric emissions (which acidify the water) – may also compound matters.

At one extreme, gross pollution episodes can result in large-scale fish kills and major changes in invertebrate populations. Indeed, the fish life of whole stretches of rivers can be effectively killed off by large farm-pollution incidents. Such pollution shows up immediately, but a single episode, even a large one, will generally have less impact in the long term than smaller but persistent discharges. Leakage over long periods can lead to a chronic lowering of oxygen levels and a permanently increased nutrient load, with more or less permanent effects on fish, invertebrate and river-plant communities. The impoverishment of the fauna and flora of chronically polluted watercourses may not be evident except to the trained observer, but where the load is sufficient there will often be obvious indications of contamination such as the discolouration of water, the presence of sewage fungus and mats of blanket weed.

Conclusions

The nature of the pollution problem from dairy farming means that, for the farmer, the problem is not so much one of causing pollution as being *detected* of doing so. The escape of effluents alone will not necessarily result in penalties being imposed. Farmers are prosecuted for causing a water pollution incident, not for the leaky effluent systems, slipshod practices or lack of spreading area that may lead to it.

Because of difficulties of detection, the full extent of dairy farm pollution and its overall contribution to reduced water quality are unknown. Certainly, by the late 1980s, farming accounted for around a third of all water pollution incidents and the great majority of farm pollution incidents are attributable to dairy farming. However, it is impossible to know how much effluent pollution these figures capture, for although the incident statistics probably include a significant proportion of the discharges that cause gross pollution, chronic low-level pollution may well go unnoticed and unrecorded. Moreover, although gross discharges are acknowledged to have serious short-term effects on watercourses, the impact of persistent low-level contamination is more contentious. The early 1980s saw a decline in river quality nationally (DOE/Welsh Office 1986a, National Rivers Authority 1991), but the extent of the contribution of diffuse farm pollution is uncertain because of the difficulty of establishing direct links between the two. Evidence supporting a link comes from detailed catchment studies such as that of the Eastern Cleddau in Wales (Schofield et al. 1990), but the sort of chronic diffuse pollution identified by such intensive investigations does not show up on routine monitoring of river quality.

Even if the full extent remains uncertain, it is now accepted that farm effluent is a significant source of water pollution. Undoubtedly, this has been the case for some years, but only in recent years has it become recognized as a public problem. The following two chapters seek to explore why farm pollution was overlooked for so long and what brought it to public recognition.

Notes

1. In a survey of 250 Devon farms, whereas 58 per cent of all the farm households derived at least three-quarters of their total household income from agriculture, some 86 per cent of those with dairy farms fell into this "most agriculturally dependent" category. These statistics are based on our own analysis of the data compiled at the Centre for Rural Studies at the Royal Agricultural College and collected in 1991 as part of a large, pan-European study of multiple job holding among farm households coordinated by the Arkleton Trust (see MacKinnon et al. 1991, Hawkins et al. 1993).

2. A survey of dairy farmers in Devon conducted two years after the introduction of quotas found that, although the overall expansion in milk production had

ended, one-third of farms had still been able to increase their production above 1983 levels (Halliday 1987; 1988). This was mainly due to the buying or leasing of additional milk quota, preponderantly by larger farms.

3. For example, the average milk yields per dairy cow in England and Wales rose from 4800 litres in 1981 to 5115 litres in 1991, an increase of 6.6 per cent (Federation of United Kingdom Milk Marketing Boards 1991: 40).

CHAPTER THREE

Farm pollution as a non-issue

Until recently, pollution from farm effluents has remained largely unregulated. This seeming laxity is partly because historically farm pollution was more localized and less severe in its consequences than it is now; but also it has only recently been recognized publicly as a problem needing regulation. Indeed, the pollution control authorities used to turn a blind eye to problems of the disposal of farm effluents.

The extension of the regulatory framework from the 1950s onwards to embrace farm pollution proceeded hand in hand with the developing recognition of the problem. However, this was by no means a linear progression. Instead, various groups successively sought to shape the definition of the problem. The initial response of law-makers and regulators was to draw farm pollution into the existing system for regulating sewage and industrial discharges, with the implicit assumption that it was no different in character. However, with the prospect in the 1970s of that system itself being tightened, the agricultural establishment successfully sought specific exemptions for farming, arguing that farm wastes were fundamentally different in character from industrial wastes and that any problems were short-term, localized and technical matters best dealt with internally within the sector. In consequence, what had been an emergent problem became something of a "non-problem" in the 1970s (Ward et al. 1995a,b), effectively internalized within the agricultural community and given exceptional legal status. However, by the mid-1980s, it was proving impossible to contain the issue in this way – with evidence of rising farm pollution incidents, a decline in river quality and in the highly charged political context following the government's proposal to privatize the water industry. For the first time, pollution from farm effluents became a public issue and was identified as a major cause of the decline in water quality.

Very quickly it was brought within a detailed regulatory framework. This chapter charts these shifts in problem definition as reflected in changes in regulatory policy towards farm pollution.

Tackling an emergent problem

The discharge of any effluents into watercourses has always been covered by common law, particularly in regard to the rights of downstream riparian owners (the owners of land bordering the riverbank). If their interests in the use of the water in the stream were damaged by pollution, then they could seek legal redress. However, starting with the Public Health Act of 1848 and the River Pollution Prevention Act of 1876, a complex body of statutory law built up for the administrative regulation of water pollution. This framework was aimed at the treatment and disposal of sewage and industrial effluents, and farm pollution was not specifically identified as a problem. As Weller & Willetts (1977: 201) have commented:

> Farmers have always held a privileged position with respect to the law relating to pollution. The law, historically, has avoided being specific and farmers have had a considerable degree of leeway in its interpretation. Industrialists, however, have had specific standards to meet since their effluent streams invariably discharge to sewer or other water under the control of the water authorities. The farmer has access to land for effluent spreading and this is under no one specific control. Pollution from such activity is extremely difficult to prove, owing to the diffuse character of the pollution and the non-continuous effluent discharge.

Indirectly, though, pollution by farm effluents was brought within statute law, most notably through fishery protection legislation, which, in seeking to safeguard a specific resource, sought to be comprehensive in covering the potential polluting threats. Such legislation began with the Salmon Fisheries Act, 1861 which made it an offence to poison salmon waters. The Salmon and Freshwater Fisheries Act, 1923 extended the protection to all freshwater fish. Under the legislation it was an offence to discharge into any waters containing freshwater fish any matter to such an extent as to cause the water to be injurious to the fish. Successful prosecutions were mounted under this Act, following fish kills from silage and slurry pollution. The Act was not a pollution control measure as such, but a prosecution measure where there had been a proven fish mortality. Even in this regard, however, it was hedged with qualifications that limited its effectiveness. The legislation was administered by fisheries protection officials and so it tended to be they, rather than pollution control officials, who acquired any experience in dealing with agricultural pollution.

The immediate post-war period saw an effort to rationalize the piecemeal structure of water pollution administration and law that had evolved since the mid-Victorian period. A key development was the establishment in 1948 of river boards with comprehensive responsibility for river pollution control. With administrative boundaries defined by catchments, the new boards assumed the functions previously exercised by a range of authorities. The substance of water pollution law was also updated in the Rivers (Prevention of Pollution) Acts of 1951 and 1961, a key provision of which was to make it an offence to use a stream for the disposal of any polluting matter. The Acts also made it illegal to discharge trade effluents into a watercourse without the consent of the appropriate river authority. By 1963, all trade discharges, including farm discharges, needed such a consent. The Public Health Act of 1961 completed the picture by bringing farm effluents under its definition of trade effluents, thus giving farmers the qualified right to require the public health authority to receive their wastes into the public sewer, while giving the authority the right to refuse consent if this was impractical or, if consent was granted, to lay down conditions, including charges.

Many river authorities took on extra staff during the 1960s to investigate the nature and extent of farm discharges into watercourses. There were thousands of applications for discharge consents; many related to septic tank outflows or runoff from manure middens and collecting yards into ditches. By mid-1969, however, of 131 171 discharges recorded from farm to river, only 3 per cent were under consent, 22 per cent were pending consent and the remaining 75 per cent were technically illegal (Gowan 1972: 26). Most authorities were overwhelmed by the number of applications for discharge consents. In Yorkshire, for example, "only those causing gross pollution and public complaint were dealt with, usually by consent refusals, to take effect over a negotiated period of time. This still left thousands of unresolved applications gathering dust" (Beck 1989: 467).

Some authorities attempted to deal with all their applications, either by consent or refusal, with varying degrees of success. However, most farm discharges, including many insignificant ones, could not comply with the then normal standard limit for any effluent, of 20 parts per million (ppm) for BOD and 30 ppm for suspended solids, as laid down by the Royal Commission on Sewage Disposal in 1912. A spokesman for one river authority was quoted as saying that the authorities imposed "realistic conditions such as 300/300 [ppm of BOD/ppm of suspended solids] that can be enforced, rather than 20/30 that the farmer hasn't a hope in hell of achieving" (Weller & Willetts 1977: 29). Likewise, although a river authority was entitled by law to sample at the point of discharge into a watercourse, many stretched the definition of effluent sampling to mean the receiving stream or ditch at the point where it left the farm boundary. This allowed a farmer to take advantage of the ditches and streams on the farm for the dilution, self-purification and re-aeration of any effluent.

With the law applied in such a lax manner, it is not surprising that it was little understood. In the words of one contemporary commentator "The farmer for his part often has no conception whatsoever of what he is doing to a stream or river or lake, and probably has no understanding of the words BOD, and indeed may also be ignorant of the basic law." (Gowan 1972: 22). In such a context, the pollution officers of the river authorities lacked legitimacy and authority, and the same commentator reported the resentment of farmers at "what they class as the snooping activities on the part of the river authority" (ibid.: 21–2). This lack of legitimacy, in turn, militated against the enforcement of the law except in exceptional cases. Even so, in 1972, 555 formal warnings were issued and 148 prosecutions brought against farmers by river authorities (Advisory Council for Agriculture and Horticulture 1975: 48). Indeed, 45 per cent of all those prosecuted for river pollution were farmers, although the average fine they faced was only £30 (Storey 1977).

Agricultural exceptionalism

Given the traditionally indulgent treatment of farmers, agricultural leaders became alarmed as pollution became a public issue in the late 1960s and early 1970s. Although attention focused on industrial and sewage pollution, there was concern lest any general clampdown might adversely affect the economic viability of agriculture and bring unwelcome interference into farming practices. Sir Nigel Strutt, the chairman of the statutory

Advisory Council for Agriculture and Horticulture, warned the farming industry:

> We are living today in the initial years of a great public movement of resentment against all forms of pollution, a movement which in the United States has reached a stage of almost hysterical crescendo. No section of the community can afford to ignore this. Agriculture is right in the thick of it. (Gowan 1972: foreword)

The following year, Strutt's Advisory Council was asked by the Minister of Agriculture to report on agricultural pollution. In addition, it was becoming apparent to some of those involved in implementing the river pollution legislation of the 1960s that the prevailing regulatory model – of discharge consents to watercourses for treated wastes or waste flows with little polluting content, and the use of the public sewerage for other wastes – was not appropriate to agriculture. It was often difficult to locate discharges from farm premises because these were rarely point discharges and were not continuous. For the same reasons it was even more difficult to monitor them. With the prospect of a mammoth task just to bring myriad farm discharges under consent, it was becoming quite evident that most of them could not possibly meet the standard conditions of consent (formulated by the Royal Commission). As Weller & Willets observed, "agricultural wastes have extremely high BOD and suspended solids levels, and discharge to watercourses is not compatible with the [Royal Commission] standards" (Weller & Willets 1977: 28).

For some farms, especially those close to built-up areas, there was the possibility of discharge to the public sewers and, particularly for many intensive livestock units with limited land, this was the only option to avoid causing serious pollution. This practice was considered by an official Working Party on Sewage Disposal which sat in 1969. In its report, the Working Party came to the conclusion that it would be quite impracticable and uneconomic for adoption on a national scale: the capital investment required at sewage works for the treatment of livestock effluents would be much greater per animal than per head of the human population. The report therefore recommended that effluents should be returned to the land wherever possible, and this became a theme that subsequent committees and reports echoed, reflecting a general unwillingness to accept farm wastes as pollutants.

The theme was picked up, for example, by the new Royal Commission on Environmental Pollution (RCEP), which was appointed in February

1970. With a membership made up of distinguished scientists and other public figures, the Commission was intended to provide independent and authoritative advice concerning the significance of, and possible solutions to, existing and prospective pollution problems. Its first report presented a broad overview of environmental problems. Under the subheading of "Pollution and agriculture", farm animal excreta was identified as a source of potentially polluting materials. The Royal Commission concluded that the problem was essentially one of "valuable manure from intensive farming [being] wasted" (RCEP 1971: para. 127). Noting that MAFF kept the issue under review and offered advice to farmers, and that additional resources had been provided for research into the matter, the Royal Commission allocated the problem to the category of "priorities for enquiry which seem to be receiving attention elsewhere and which, despite their importance, we do not propose to tackle ourselves, though we shall be concerned to see prompt and effective decisions" (ibid.: para. 130). In its fourth report (RCEP: 1974), the Commission indeed returned to the problem, to urge that a continuing effort be made to improve storage facilities for excreta on farms so that material could be spread on land at the optimum times to supply crop needs and minimize pollution risks.

The report of the Agricultural Advisory Council

This pressure from the Royal Commission and its implied threat of independent investigation may well have been a factor in prompting MAFF to ask its own Advisory Council for Agriculture and Horticulture (ACAH), under the chairmanship of Sir Nigel Strutt, to undertake a review of problems to do with farm wastes. The government was also committed to the introduction of legislation that was to consolidate and strengthen controls over water pollution, and there was concern within the Ministry, as well as within the farming industry, over what the implications would be for agriculture. In February 1973, therefore, MAFF requested its Advisory Council:

> to investigate and report on the extent to which agriculture contributes to pollution through the use of fertilizers and by the disposal of farm waste; whether control measures are adequate or unnecessarily restricting and any action which may be desirable. (ACAH 1975: 1)

The Council was composed of distinguished agriculturalists and horticulturalists, together with a senior MAFF civil servant. It was the Minister of Agriculture's official advisory body on broad questions of agricultural policy. Over the following 22 months it produced three reports: on pollution from nitrogenous fertilizers, on farm odours, and on the disposal of farm wastes. Its report on farm wastes concentrated on the water pollution problems arising from livestock manures and silage effluent. It was at pains to distinguish the circumstances of farm pollution from those of industrial pollution, pointing out that:

> agriculture, unlike other industries, is in a special position in that the land, which is agriculture's raw material, is also a source of water. Excessive restrictions placed on the agriculture industry with the aim of improving water supplies can only be at the expense of domestic food production. (ACAH 1975: 5)

In other words, a balance had to be struck between the competing needs of food and water supply. The Advisory Council went on to challenge the applicability to agriculture of the "polluter pays" principle that had recently been adumbrated by the Royal Commission on Environmental Pollution and the OECD and assented to by the EC Council of Ministers. The Committee argued that, agriculture was a "price taker" rather than a "price setter" and it was not possible for it to pass on the costs of pollution control measures automatically to its customers, with the consequence that "a farmer might be required to bear the cost of control measures imposed on him, or have restrictions imposed on his activities for, say, social reasons when in terms of good agricultural practice these measures are not necessary" (ibid.: 28).

The Advisory Council also reasoned that farming was not intrinsically a polluting activity in that the problems of farm wastes "are not primarily ones of disposal but rather of better resource utilization and standards of management" (ibid.: 2). The fundamental problem, therefore, was where holdings had insufficient land for proper utilization of effluents. This seemed to be specific to the intensive pig sector. "Nearly all dairy enterprises appear to have sufficient land to spread their manure" (ibid.: 4) – the June 1973 farm census indicated that there were only 200 dairy holdings keeping cows at densities in excess of 2.5 cows per acre (6.2 per hectare), and most of them were the survivors of the old town dairies. Somewhat disingenuously, the report pointed to the few formal warnings and prosecutions against farmers for causing water pollution as an indication of the exceptional nature of the problem, while commending "the attitude of most authorities in seeking to persuade and educate rather than resorting to formal legal sanctions" (ibid.: 12). Still, only a very small proportion of farmers seemed to be causing serious offence.

The Advisory Council's report conceded that "agriculture is no doubt a source of a certain amount of diffuse pollution" (ibid.: 12), but such problems were minor and should in any case largely solve themselves. Inorganic fertilizers, which were no longer subsidized by the Ministry, had risen markedly in price in recent years. In consequence, "livestock manures are now an increasingly valuable resource and the cost savings to be achieved by good use of the nutrients have become an important consideration" (ibid.: 2). Many of the practical precautions which farmers could take to prevent pollution were already spelt out in advisory leaflets issued by the Ministry. Any shortcomings therefore were essentially ones of information and training: "Farming expertise in agricultural waste management and pollution control", declared the Advisory Council, "is significantly less than in other sectors of farming" (ibid.: 6). Thus, many of its recommendations related to enhancing ADAS's expertise in the management of farm wastes; publicizing the advice ADAS provided on pollution prevention; encouraging farmers to seek such advice; drawing particular pollution risks to the attention of the farming community; and improving liaison with planning and water authorities.

The Advisory Council, however, clearly felt that much greater sources of ignorance, outside of farming, also needed to be tackled and that these misperceptions were as much part of the problem as were unfortunate cases of the runoff of farm effluents. It called for improvement therefore to "the education of the general public in the ways of the countryside, especially those who have gone from the towns to live in the country, as well as officials of public bodies, such as those concerned with health and water supplies" (ibid.: 30). Not only should the public be given a better understanding of the problems faced by farmers, but it should also be made more aware of its dependence on the farming industry for much of its food supplies and the threat that restrictions might pose of reduced production. Indeed, the Council urged the need to safeguard agricultural interests from any future restrictive measures that might threaten a drop in farm output, and this is exactly what the government did.

The 1974 Control of Pollution Act
The Advisory Council's deliberations on farm pollution coincided with the passage of the Control of Pollution Act 1974 (COPA). Although this

legislation promised potentially new controls over serious sources of pollution, it also reflected the sort of defensiveness and complacency towards agriculture that the Advisory Council displayed. Thus, on the one hand, it strengthened the duties and power of water authorities both to forestall and remedy water pollution. The greatest potential implications for agriculture arose from two changes. First, the offence of causing water pollution was extended to include underground water with gathering evidence of nitrate contamination of some groundwater sources. Secondly, a reserve power was introduced whereby the Secretary of State for the Environment could designate as a water protection zone an area where there was a risk of water pollution and, within it, could proscribe activities that could not then be carried out without the consent of the water authorities. On the other hand, the legislation gave a general exemption from prosecution for causing water pollution to any farmer operating in accordance with "good agricultural practice".

A Department of the Environment (DOE) guide to pollution control in Great Britain summarized the reason for the provision as being that: "normal farming practices may sometimes result in water pollution and farmers are therefore afforded some protection by the Act" (DOE 1976: para. 84). The Advisory Council for Agriculture and Horticulture explained the exemption in the following terms:

Under this Act the offence of causing or knowingly permitting the contamination of water is extended to include underground water . . . In recognition of the possible difficulties which these provisions could cause for some farmers, particularly as regards underground water, special provision has been made for bona fide agricultural activities. (ACAH 1975: 23)

The exemption caused hardly any contention during the Bill's parliamentary passage. On the contrary, grave misgivings were expressed by MPs and peers with agricultural connections at the possible threat posed to normal farming practices by the all-embracing nature of the offence of causing water pollution, as well as the reserve powers of blanket regulation in water protection zones. While ministers gave assurances that these powers would be used sensitively and sparingly (the powers, in fact, were never used), they resisted calls for farmers to be compensated for restrictions on their activities, even if such restrictions ran counter to good agricultural practice.

Nevertheless, the Act contained crucial safeguards that preserved the autonomy of the agricultural sector against creeping pollution controls. The most crucial was the "good agricultural practice" defence. The legislation left it to MAFF to specify in a code what should count as good agricultural practice, and thus to determine which normal farming activities should transcend pollution controls. However, a water authority could apply to the Secretary of State for the Environment for a notice to request a farmer to desist from such an activity, if it was thought to be causing pollution, and this would then open the farmer to prosecution if pollution subsequently occurred.

Special arrangements were also made for agriculture in establishing water protection zones. These allowed, first, for an agricultural assessor to assist an inspector at the public enquiry, which would normally be held before regulations were made. Secondly, where farmers wished to appeal against the refusal of a consent for a proscribed activity, they would be able to apply to the Minister of Agriculture for a certificate of good agricultural practice in respect of that activity; and if a certificate was granted the appeal would be determined jointly by the minister and the Secretary of State for the Environment.

Pressed to say why such exemptions should not apply to other sectors, such as quarrying, an Environment Minister responded:

> The special provision in the bill has been limited to farmers for three good reasons. First of all, modern farming involves applying extensively to the soil substances which, in sufficient concentrations, would pollute water unacceptably. Secondly, agriculture still uses so much of the surface of the land ... that in the aggregate farming is far more at risk than any other industry of inadvertently polluting water. Thirdly, it is still an industry in which a large number of people work in small units, on their own account. (Lord Sandford, Hansard, House of Lords Debates. 22 January 1974 col. 1424)

Among contemporary commentators the provision did not seem overly tendentious. For example, a leading commentator on environmental law remarked on this "concession to farmers", in the following terms:

> the idea behind the provision must be that the farmer will be permitted to carry on with normal farming practices until the resulting pollution is such that, on the balance of advantages to the community,

it is better that he should change his methods. The discretion will lie with the Secretary of State who will have power to issue the notices. The procedure will be characteristic of English pollution legislation. (McLoughlin 1975: 78)

The water industry seemingly put up no concerted resistance. Nevertheless, there was significant concern internally about agricultural pollution, as revealed in the earlier efforts of the river boards to record and register farm discharges, the rising number of farm pollution incidents mentioned in their annual reports, the level of formal warnings and prosecutions issued against farmers (555 and 148 respectively in 1972), and a spate of papers on farm wastes in the *Journal of the Institute of Water Pollution Control* in the mid- to late 1970s.

However, the water industry's highly fragmented structure did not lend itself to effective lobbying or give it a coherent voice, and it was in any case being subjected to a periodic bout of reorganization. The 1973 Water Act replaced the 29 river authorities, 157 water undertakings and 1393 sewerage units operating in England and Wales with just ten large regional organizations responsible for all aspects of water, from source to sewer. Although these regional water authorities would become more powerful organizations, the passage of the Control of Pollution Act 1974 coincided with the highly disruptive period of their establishment. Subsequently, though, the new water authorities expressed to the Royal Commission on Environmental Pollution their concern that "the provisions of the Act may be held to place agriculture in a special position in relation to pollution" and "that, in conflicts between agricultural and pollution considerations, the former may be regarded as paramount" (RCEP 1979: 3).

With an overriding policy commitment to expand food production in Britain, there was little room for the recognition of farm pollution as other than a very minor and peripheral concern. This commitment was actually strengthened in the mid-1970s. Britain's entry into the EC and its Common Agricultural Policy coincided with concerns about possible world food shortages, which led to a renewed imperative for agricultural expansion. The 1975 White Paper on agriculture, for example, even though setting out a target for expansion of 2 per cent a year, which was bound to involve considerable intensification of production, made no more than a passing reference to pollution (MAFF 1975).

The view within MAFF was that pollution problems were ephemeral and that technical solutions would be found to any short-term difficulties in the handling of farm wastes. A paper prepared by ministry staff, "Agriculture in 2000 AD", predicted a continuing trend in livestock farming towards higher stocking densities, larger units and the housing of animals (MAFF 1979). Greater quantities of effluent would need to be disposed of, but the paper was optimistic about "further developments in effluent treatment and handling, including improved usage as manure or in recycled form as animal food" (ibid.: 248). All new and modernized dairy units were expected to adopt cubicle housing, resulting in an increase in slurry handling, but environmental problems were expected to decline, with the development of techniques of applying slurry to land and separating the solid and liquid fractions, to the point where the nuisance might be "no greater than from conventional manure handling systems involving [farm-yard manure]" (ibid.: 252). At the same time, improved hay-making methods were thought likely to halt or even reverse the trend towards more silage, thus reducing the risks of pollution from silage effluent.

The significance that MAFF accorded farm pollution at this time can be gauged from the fact that, out of an ADAS staff complement of 5800, the Farm Waste Unit comprised just four specialists and one clerk. Their function was to act as a source of expertise and guidance for local agricultural advisory staff, but the location of the Unit in the Agricultural Science Service, rather than in one of the "front line" advisory services, indicated the way in which farm pollution was regarded as a minor, technical matter, detached from the central concerns of crop and animal production and farm management. In keeping with this outlook, greater resources were allocated to research: in 1976 MAFF devoted £0.25 million pounds to work on farm wastes; and the Agricultural Research Council and Department of Agriculture and Fisheries in Scotland also supported research in this field. The Agricultural Research Council posed the problem in the following terms: "In nearly all circumstances in the United Kingdom slurry produced in livestock units is sooner or later returned to the land . . . Its treatment and effective use are therefore primarily agricultural problems" (ibid.: 131).

The 1979 Royal Commission report

In 1977 the Royal Commission on Environmental Pollution launched an investigation of agriculture and pollution, which emerged as its seventh report in 1979. This was only the second time it had taken a detailed look at a specific sector, the previous one having been nuclear power.

Whereas the investigation into nuclear power had covered an issue of great contention, agriculture was a sector that at the time aroused little controversy regarding pollution. Evidently the Commission was goaded into action by the widespread complacency it found in the agricultural sector, rather than by specific problems. In a preliminary discussion held with MAFF the Commission members "were not persuaded that sufficient attention was being paid to the pollution that might be caused by agriculture . . . we were left with the impression that such problems were regarded as secondary in importance and as unavoidable concomitants of food supply" (RCEP 1979: 3). The Commission took an inclusive view of agriculture and pollution, covering not only pesticides, nitrogen fertilizers and farm effluents, but also the effects of pollution from urban and industrial sources on agriculture itself, in obeisance to the widespread view within the sector that farming was more sinned against than sinning.

In relation to farm wastes, the Commission emphasized how the 1975 report from the Advisory Council for Agriculture and Horticulture had provided a valuable starting point and assistance to its own inquiries. The categories of waste covered included manures and slurries, cereal straw, silage effluent, sheep-dip liquors and surplus pesticides and pesticide containers, although the bulk of attention was devoted to pollution risks from slurry, which the Commission regarded as the major problem. On some of the other problems, the Commission took what proved to be an over-optimistic view of the prospects, for example, expressing itself "satisfied that in view of the extensive advice that is now made available on silage-making, and the increased awareness by farmers of the risks, pollution from this cause should be a diminishing problem" (ibid.: 154).

Regarding animal effluent, the Commission drew a basic distinction between "traditional, mixed farming systems [in which] the manure produced is part of a balanced, self-sustaining cycle, in which nutrients are returned to the land" and large-scale intensive livestock units established on relatively small areas of land and in which "animal production is virtually a factory process for converting grass or grain into meat or eggs" (ibid.: 127). In the first case, the Commission questioned the appropriateness of the term "waste": "given good management, the storage and spreading of the manure presents no significant pollution problems" (ibid.: 128); moreover, it had considerable nutrient value as a fertilizer and was a good soil conditioner. In the second case, though, where "the area of land available for the disposal of excreta may be far less than that required for the nutrient value of this material to be efficiently used . . . the excreta

51

may indeed be regarded as waste; that is, as something of little intrinsic value, to be disposed of as cheaply as possible" (ibid.: 128). For this reason and because of the much higher pollution risks involved, the Commission judged that "intensive livestock units are *not intrinsically agricultural* in character; they are *essentially industrial* enterprises and should be regarded as such" (emphasis added). The implication was that, "as with other industries, the need for pollution control, and the need to bear the costs of that control, must be accepted" (ibid.: 128). The Commission, therefore, recommended that from the perspective of both planning control and pollution control, the development and operation of intensive livestock units should be subject to local authority regulation, as well as MAFF inspection.

The Commission's distinction can be seen to be an essentialist one, contrasting livestock farming proper, to which pollution should be an anathema, with factory production of animals, whose high polluting risks confirmed its separate industrial character. The difficulty with the distinction was that it left unclear the status of specialized animal production, particularly dairy farming, which had not undergone the processes of concentration to the same extent as had occurred with pigs and poultry, but which was increasingly employing intensive practices. However, the Royal Commission was preoccupied with large intensive pig units. Its members paid a visit to North Humberside, where there was the greatest concentration of such units in Britain, causing locally acute problems of pollution and nuisance, which evidently left a strong impression on them. As their report remarked "The smell of pig slurry, as we experienced when visiting Humberside with its many pig units, is highly offensive and penetrating" (ibid.: 134).

However, the Commission could not fail to acknowledge adverse trends in dairying, including a growth in herd sizes, an increase in the number of large herds on relatively small farms, and a strong trend towards slurry-based housing systems, all of which led to "the risk of pollution from leakage or overflow, from the runoff of slurry, and of smell from that slurry spread on land or stored" (ibid.: 17). Nevertheless, its overall judgement was that "The pollution problems ... appear mainly to be down to pig and poultry units rather than to cattle units" (ibid.: 133). Indeed, in proposing a definition of an intensive livestock unit, based on its polluting potential, for the purposes of statutory regulation, the Commission suggested that it should not include the winter housing of cattle (ibid.: 193).

Because of the concentration and heavy regional specialization that had occurred in pig and poultry production, the Commission remarked that the pollution problems associated with the storage and disposal of animal excreta were localized and usually sporadic. They identified three types of pollution risk: water pollution, transmission of disease and smell. For the first two, the Commission looked to technical solutions. Water pollution risks could be avoided or minimized by care in the siting and construction of waste storage facilities and by conscientious management of these facilities and subsequent spreading operations. Likewise, the risk of infection would be minimized by not spreading onto grazing land or by storing the slurry for at least one month and then leaving the land ungrazed for at least six weeks. By far the most intractable problem was that of the smell. As the Commission commented "It is clear that the smells associated with intensive animal production give the most offence to, and cause most complaint by, the public at large. It is also the most difficult problem to define" (ibid.: 134).

Although the Royal Commission proposed that intensive livestock units should be subject to the planning laws, the whole thrust of its report was that farm pollution should primarily be the concern of the agricultural sector itself. Its report called for "a new emphasis on pollution matters within MAFF", and the Ministry was criticized for having been in the past "unduly defensive and protective towards agricultural interests". The Commission was adamant that "the initiative for action to deal with pollution caused by agriculture should rest with MAFF" (ibid.: 211).

Farm pollution regulation in the 1970s and early 1980s

The Control of Pollution Act 1974 (COPA) had left the objectives of farm pollution control ambiguous. All that was clear was that the pursuit of water quality had to be qualified by considerations of agricultural efficiency and productivity. Unclear objectives were compounded by a dearth of information on the problem. Prior to the Act, the level of prosecutions of farmers had been one indicator of the scale of serious farm pollution. However, after 1973 the number of prosecutions dropped sharply and did not reach their pre-COPA levels again until the mid-1980s (see Table 3.1). Significantly, the number of written warnings fell much less sharply than

Table 3.1 Number of convictions for farm
pollution (1972–91)

Year	Number of convictions
1972	148
1973	69
1974/5	23
1976	n/a
1977	28
1978	n/a
1979	38
1980	34
1981	71
1982	64
1983	87
1984	110
1985	159
1986	128
1987	225
1988	148
1989	163
1990	221
1991	214

Sources: Advisory Council for Agriculture and
Horticulture (1975), Grundey (1980), NRA (1992b).

the number of prosecutions, indicating a more marked reluctance to pursue
cases through the courts. After 148 prosecutions in 1972, there were just
23 in 1974 and 28 in 1977. The drop was not solely due to the exemption
for farmers provided by COPA (the relevant section of which did not, in
any case, formally come into force until 1986). The legislation had coincided with the reorganization of the water industry in which the river
boards were subsumed into the multipurpose regional water authorities.
For the former, river quality had been a prime objective, but this became
secondary to the water authorities' overriding concern with water supply
and sewage treatment.

   Whatever the cause, farmer prosecutions were clearly no longer an indication of the scale of the problem (if they ever had been). During the 1970s,
indeed, water pollution from farming became something of a non-problem.

The water authorities acquiesced in the priority given by government to the expansion of food production over the curbing of farm pollution. This led to a lack of effort in monitoring and measuring the problem. Schemes initiated by the former river boards, to bring all significant farm discharges under formal consent procedures, were no longer pursued. The main river monitoring activities of the water authorities focused on major stretches affected by urban and industrial pollution and neglected the rural catchments where farm pollution might be prevalent. In any case, the periodic sampling procedures used were not designed to pick up the diffuse or spasmodic pollution typically caused by farming activities. Hence, there was a lack of information about the extent of the farm pollution problem, which completed a vicious circle that, in effect, closed off the issue. For the dearth of information meant, in turn, a lack of public awareness of the problem, with little pressure on water authorities to do anything about it, or on politicians to acknowledge its existence.

On the surface, all seemed well. Water authorities were reported to believe that "negotiation, rather than punitive measures, will lead to better cooperation" (Weller & Willetts 1977: 29); and a National Farmers' Union spokesman remarked that the authorities had been "taking a cooperative and helpful line over discharges of farm effluent and certainly few cases come to us for attention" (quoted in Weller & Willetts 1977: 29). However, little regulation of farm waste pollution took place, and the submerged reality of what did occur was much more problematic, as research by Keith Hawkins at Oxford University subsequently revealed. Reporting on participant observation fieldwork with water authority pollution control staff, Hawkins remarked:

Farmers are an occupational group regarded by officers with rural patches as particularly troublesome, partly because of lack of resources, partly because of a characteristic stubbornness born of decades of water-use unencumbered by the attentions of any regulatory agency. They have a culture of their own which sometimes impedes the officer in doing his job: "They're a tight-knit community," said an officer of long standing from a large rural area, "and they won't inform against one another. There's a lot of cover-up." Indeed, of all the occupations regularly encountered, the farmer is the most consistently described as difficult to handle, as an area supervisor's question, posed with deliberate ingenuousness, acknowledges: "Farmers, in particular where you get an uncooperative one, are [some] of the worst customers that we come across. I don't know whether you've heard that before?" (Hawkins 1984: 115)

Dealing with farmers was seen to demand particular tact and diplomacy, and Hawkins found that pollution control officers in rural districts learned early on that it was essential, in the words of one of them, to "avoid getting their backs up. You get their backs up and you'll get nowhere". Too ready a reference to the law was considered to be counterproductive, as another of them explained, in describing how he approached farmers:

I try and take an interest . . . in the farm, show them – try and show them – I'm not just someone who comes along from an office in his suit who doesn't know anything about farming at all. I may not approach the question directly. I may have a chat about the farm, see what's he's got . . . You go along and be bloody-minded to a farmer, try and lay the law down, go quoting this Act and that Act which he's probably never heard of and doesn't understand, he'll just put his foot down and you won't get anywhere. (ibid.: 136)

Given the difficulties in dealing with farmers and with farm pollution, field staff were inclined to turn a "blind eye" to the problem as long as there were no wider repercussions. Such an approach was often condoned by their area supervisors, one of whom was quoted as saying he would "bend the rules on farms", because: "farmers are one of the most intractable problems you can get. You can't treat farm effluent satisfactorily . . . So far as I'm concerned the farmers can continue until someone squeals" (ibid.: 217).

Farm pollution thus went unregulated in practice. Official neglect was matched by and helped perpetuate public unawareness, which in turn fostered political indifference. The period from the late 1970s to the mid-1980s was indeed a political hiatus for problems of agricultural pollution. The Royal Commission had deliberated on the subject and had produced a report which, although critical of the agricultural sector, had looked to it to put its own house in order. The Commission had made various proposals for improved practice and administration, relating not only to intensive animal husbandry but also to the use of pesticides and nitrogenous fertilizers. Even so, the government took four years to respond (DOE 1983) and then rejected some of the key proposals, including a requirement for agricultural capital grants to be made conditional upon water authority consultation. In principle, COPA offered certain additional controls, but these too were subject to government delay and equivocation. The provisions of this Act relating to water pollution (Part II) did not start to come into effect until July 1984, ten years after the legislation had been passed. The farming lobby was particularly concerned about the extension of water authority powers to control discharges to land, and the National Farmers' Union and Country Landowners' Association, with the assistance of MAFF, negotiated an agreement which specifically exempted farmers from the operation of the proposed regulation.

In February 1984 MAFF circulated a draft of the *Code of Good Agricultural Practice*, as required under Part II of COPA. The water authorities reacted strongly. They were reported to feel that "many of MAFF's advisory leaflets which are to underpin the code are so overwhelmingly concerned with agricultural productivity as to be virtually useless for water control purposes" (ENDS 1985: 21). In response, the Code, finally issued in January 1985, was substantially rewritten and stressed – unlike the draft – that "in planning and operating a farm enterprise it is essential to take precautions against all forms of pollution" (MAFF 1985). Although the Code had been intended to be a defence of responsible farmers, and the farming and landowning lobbies had been consulted on it, there was unease among them at its declaration that "nothing that is likely to cause water pollution could be accepted as good agricultural practice". MAFF responded by drawing a commitment from the DOE to write to all water authorities underlining the need to interpret the Code in a reasonable manner and committing itself to monitor the administration of the Act and to "take the necessary action" if the water authorities proved unaccommodating (*Country Landowner*, April 1985: 9).

Conclusions

Thus, throughout the 1970s and early 1980s, agricultural pollution as an issue was very much under the control of the agricultural community. As such, it was largely removed from political debate and, with the exception of the smell arising from intensive livestock units, aroused little, and then only localized, public concern. The response of the agriculture departments was equally low key and consisted almost wholly of the funding of research and development, much of which was directed at odour control.

Information about the subject was available to farmers through ADAS, and grant aid could be obtained for waste facilities installed under farm development schemes. There were, however, no regulations governing the storage or disposal of farm effluents, and the subject had a low profile, so there was little incentive for the farmer to take action to prevent pollution. On the contrary, the much more powerful stimulus of production policy acted to increase pollution risks, not only indirectly through its encouragement to intensify production, but also directly through the provision of grant aid to such measures as the concreting of yards, the construction of animal housing and the under-drainage of fields. As a water official in the South West commented subsequently, "the ministry . . . dumped large amounts of money into [dairy] farms to change their practices. It never seemed to occur to anyone to tell them what to do with the pollution" (*The Independent,* 16 October 1987).

Retrospectively, as we shall see, it became evident that pollution from livestock farms was widespread and systematic. MAFF's own surveys revealed that, on many farms, livestock wastes were being applied at rates higher than their take-up by crops; and in the South West of England in particular significant amounts of dairy cow manure were being applied at rates far in excess of crop requirements (Richardson 1976). What helped keep such a chronic and endemic problem from wider recognition was that most of it was low level and spasmodic and it occurred away from major abstraction points. But the acquiescence of many farmers, agricultural advisers and regulatory officials in this state of affairs points to the importance also of the prevailing conceptual categories through which the issues were understood and through which fundamental boundaries were maintained.

As we have seen, various sources of authority had drawn upon and reinforced a fundamental distinction between agriculture and industry in reviewing the issue of farm waste. None of these was blind to the fact that farming was changing in its scale and intensity. As the Royal Commission commented, "Farming, long regarded as a way of life, is now becoming as much an industry as many other businesses" (RCEP 1979: 2). It was generally acknowledged, moreover, that these changes were generating increased environmental pressures. Even so, although ministers and the Agricultural Advisory Council addressed various arguments to justify its exceptional treatment in environmental law, the distinctive character of agriculture remained something unquestioned – a fundamental assumption that required no explanation precisely because it went unchallenged, but which underpinned the reasoning that, because of its natural origins, farms produced "waste" but not "pollution". Only the Royal Commission felt the need to reflect briefly on what constituted agriculture's fundamental distinctiveness and it identified this in agriculture's dependency on "the weather, type of soil and other variables that make farming an art as well as a science" (RCEP 1979: 128). In a move revealing of the wider issues at stake, the Commission proposed to move the boundaries, in effect to preserve the categories: those husbandry activities that the Commission recognized as significant sources of pollution – namely, intensive livestock rearing – it redefined as not agricultural but industrial. Agricultural pollution was thus not only an aberrant act but also an anomalous category. The recognition of agricultural pollution would therefore necessitate a major shift in meanings, and we turn next to consider under what circumstances and with what consequences this shift actually occurred.

CHAPTER FOUR

The politicization of farm pollution

Introduction

In 1983, the DOE formally responded to the Royal Commission's report on Agriculture and Pollution. The 34-page response, which was given over mainly to pesticides and nitrate fertilizers, included just two-and-a-half pages on animal wastes and these made no specific reference to water pollution (DOE 1983). Instead, it simply echoed the received wisdom on the issue, namely that the problem was essentially one of intensive livestock units, mainly in the pig sector but also in the poultry sector, and that the chief concern was to deal with the smell associated with the waste. Yet again, a code of good practice was seen as the way forward. The government rejected proposals that grant aid should be linked to pollution control measures and that there should be a formal requirement for water authorities to be consulted over planning or grant aid applications for agricultural developments likely to cause water pollution.

In stark contrast, just five years later the government accepted that farm wastes were a major source of water pollution which needed to be regulated (DOE 1988). This admission was made in response to a report on river pollution by the House of Commons Environment Committee (1987). In the short intervening period between the two government statements, therefore, water pollution from farm wastes had emerged as a significant and a politicized problem. It is unlikely that its extent or scale had changed significantly. What then had happened to give farm effluents such prominence as an environmental issue?

Crucially, the two policy communities that until then had either suppressed or ignored the problem had been thrown into disarray. On the one hand, the agricultural policy community had been rocked by the EC's imposition of milk quotas in March 1984, which, at a stroke, had undermined its single-minded preoccupation with boosting production. On the other hand, government plans to privatize the water authorities, published in February 1986, had provoked considerable opposition which had engulfed the water policy community. Of all the privatizations of the Thatcher governments, this proved the most controversial, involving the transference of a very traditional and vital public service into private hands. With government so evidently willing to divest the public sector of its responsibilities, there was scope for others to assert their claims to represent the public interest, and, in particular, this provided a major opportunity for regulatory officials and pressure groups to promote the new environmental morality.

Out of the breakdown of these two important communities, new groupings emerged, as both insider and outsider groups sought to shape a very fluid policy agenda and to position themselves in relation to any reconstitution of the policy communities. One such grouping brought together both disaffected regulatory staff of the water authorities and various interest groups opposed to the plan to privatize the regulatory functions of the regional water authorities alongside their service functions. In seeking to generate support for the water authority's regulatory functions, they began to expound a rhetoric concerning the need for independent regulation in the public interest. Nothing seemed to illustrate that case more than the neglected issue of farm pollution. The turbulence of the agricultural policy community provided the opportunity for this issue to be prised from it and to be redefined as an environmental problem rather than as a technical problem of production. With the loss of the imperative to expand agricultural production, it could no longer be convincingly maintained that farming objectives should automatically take precedence over environmental considerations.

The Environment Committee's investigation of river pollution was a key event in politicizing the issue of farm pollution, as well as in formulating a consensus between agricultural and water interests over how the problem might be tackled, which was then taken up in government policy. The investigation had been triggered by the publication of the quinquennial survey of river quality conducted in 1985 and published in the following year (DOE/Welsh Office 1986a). There were reductions from the previous survey (undertaken in 1980) in the lengths of river classified at either extreme as either "good" or "bad" quality, with an associated increase in the proportion of rivers within the middle categories of "fair" and "poor".

The net shifts were slight but the overall change could only be interpreted as a deterioration, making it the first in the seven national surveys conducted since 1958 not to record an overall improvement. In interpreting the results of the river quality survey, another set of statistics, also issued in 1986, but likewise referring to 1985, was very influential – namely the compilation of reported water pollution incidents caused by farm waste in England and Wales (Water Authorities Association/MAFF 1986). It drew public attention for the first time to the widespread and seemingly growing incidence of farm pollution. These two publications together transformed perceptions of the farm waste problem. They had such an impact, partly because of the new information they contained in respect of trends in river pollution and its causes but also because of the political context in which the statistics appeared.

The politics of water privatization

What made evidence of increasing water pollution so politically charged was the announcement of the government's proposals for the privatization of the water authorities (DOE/Welsh Office 1986b). These envisaged that the existing regime for water pollution control would remain intact: the regulatory duties and functions of water authorities would simply pass to the new private water service companies. The proposals caused considerable consternation, which the publication of the river quality survey only served to fuel. The various interpretations that could be put upon the survey's findings became central to an intensifying debate over what was wrong with water pollution regulation and whether the government's privatization plans would improve or worsen matters. Those opposed to privatization could argue that the deterioration in river quality was already an indictment of the weakening of public controls and of public investment within the water industry that privatization would only serve to exacerbate. Those in favour of privatization were able to point to the apparent failure of public regulatory bodies in safeguarding the water environment. As well as this partisan debate over the rights and wrongs of Thatcherism, discussion and interpretation of river quality statistics were also joined by various groups – some within the water industry, others from outside, including environmental groups and industrial users of water

– all seeking to influence the form that privatization might take and any safeguards that the government might feel obliged to introduce.

The regional water authorities had been set up in 1973. Based on the concept of integrated river basin management (IRBM), they exercised control over every aspect of water management in their particular region. As Richardson et al. comment: "Dominated by engineers, the water policy sector conformed fairly closely to the model of a policy community in which technical expertise was the main basis for consensual decision making" (1991: 7). Environmental groups were not included: the only environmental interests represented on the boards of the water authorities tended to be MAFF appointees from Regional Fisheries Committees. Significant changes occurred in the policy sector during the early 1980s. In 1983, the last vestige of local authority control was removed with the ending of the appointment of members to the regional water authorities by local authorities. At the same time, the National Water Council, the planning body for the sector, was abolished to be replaced by a trade association, the Water Authorities Association (WAA), which quickly established a stronger corporate identity and lobbying potential for the water industry nationally. Under the impact of tighter budgetary restrictions and a climate of opinion increasingly hostile to the public sector, water came to be seen less and less as a service and more and more as a commodity, and an ethos of commercialism began to prevail (Saunders 1985).

It was within this context that the notion of water privatization emerged, being first proposed publicly by a chairman of one of the regional water authorities. Indeed, a strong driving force to privatize emanated from senior water authority management. The proposal was taken up by right-wing ministers. Plans for privatization were drawn up in consultation with the water authorities, but when published they caused such contention as to throw wide open the policy community. It was evident that many ordinary people were deeply shocked at the prospect that such a basic resource should become subject to commercial control. Richardson et al. in their study of water privatization (1991; see also Maloney & Richardson 1994, 1995) argue that the proposals for privatization and the response to them had the effect of destabilizing the existing policy community. The resulting political void, they argue, allowed many previously "outsider" groups to become at least temporarily involved in this policy field. They suggest that it was in the possible reconstitution of a new policy community that these outsiders saw their chance to influence decisions. Several hundred interest groups were aroused to lobby on the matter, and at the same time, privatization plans revealed critical fissures within the water authorities themselves.

Opinion was divided not only between water authorities but also among the staff they employed. The WAA adopted a stance that, if privatization was to go ahead, then the authorities should be privatized whole, as they stood. Likewise, the main professional body, the Institution of Water Engineers and Scientists, argued that the principle and practice of IRBM must be preserved in any changes, and that it was therefore essential to retain within a single structure the environmental management of each area, including the regulatory functions of resource management and pollution control (see Kinnersley 1988, 1994, Maloney & Richardson 1994, 1995). However, the main grouping of regulatory staff, the Institute of Water Pollution Control (IWPC) although also arguing for the retention of IRBM, felt that maintenance and control should be carried out by an independent public body (something akin to the River Purification Boards that still existed in Scotland)[1]. The IWPC thereby aligned itself with a range of groups, including environmental, industrial, landowning, fisheries and angling concerns, who, in opposition to the government and the WAA, wanted to see the retention in the public sector of the key regulatory functions. Out of this commonality of interests, one can see the beginnings of the emergence of a new policy network, of which the National Rivers Authority was the eventual incarnation.

The question arises of why the IWPC adopted this dissenting position. Its members had mainly worked for the English and Welsh river boards and had tended to lose status and independence when these boards had been subsumed into the regional water authorities. They had also suffered disproportionately in the financial squeezes and managerial reforms of the early 1980s. To maintain performance targets with reduced staffing levels, it had tended to be the regulatory and monitoring activities rather than the operational and service functions that had suffered the greatest cuts. For example, in the South West Water Authority there was a 50 per cent reduction in sampling for river quality monitoring between 1980 and 1985 (House of Commons Environment Committee 1987: 308). It was unlikely that, within privatized utilities, such priorities would be reversed; indeed, it was likely that the pressures on such unprofitable functions might intensify. Among this grouping, therefore, there was an emerging interest in a higher public profile and increased political support for public regulatory functions. The publicizing of farm pollution helped achieve just this objective.

Farm pollution incidents

Here was an apparently novel threat to the integrity of rural rivers and streams, although it had concerned water authority staff for some time. What had been lacking in the past was systematic evidence of the problem. Commenting on British pollution control practice in general in that period, one perceptive American commentator remarked, "On the one hand we are told that the system is one of the best in the world and that pollution is firmly under control. Yet there are persistent and nagging doubts. Much of what one would expect in a well developed control system is missing." (Hays 1984: 22). Traditions of administrative secrecy, pragmatism and informality, he argued, rendered the objectives of pollution control opaque and made it impossible to assess its achievements. "Are there no firm numbers?" he was led to complain (ibid.: 22). It is, after all, through the medium of statistics – expressed as rates, levels, trends, correlations and the like – that the significance and progress of public problems are usually gauged and judged (Porter 1995).

Such scepticism would have applied with particular force to the area of farm pollution. The Control of Pollution Act (COPA) had left the objectives of control ambiguous, with the pursuit of water quality qualified by considerations of agricultural efficiency and productivity. Unclear objectives were compounded by a dearth of information. As we have seen (Table 3.1, p. 54), COPA led to a sharp drop in the prosecution of farmers, which not only cut off the main source of publicity about farm pollution, but which also meant that the level of prosecutions was no longer an indication of the scale of the problem. The dearth of information in turn helped close off the issue from public debate.

Of course, individual water authorities did have information on the cases of farm pollution that they had to tackle. Eventually, the gathering, systematic collating, publishing and publicizing of statistics on such incidents were to provide the missing numbers that enabled farm pollution to become a public issue through becoming a measurable problem (WAA/MAFF 1986). Indeed, it may be argued that it was necessary to construct the notion of a "farm pollution incident" – by turning isolated, individual cases into statistical events – to break out of the vicious circle of this "non-problem" (Hill et al. 1989, Ward et al. 1995a).

But what is a farm pollution incident? It is essentially a gross pollution event whose consequences (say, in the form of a fish kill, sewage fungus or a discoloured stream) are observed and hence the event is reported; but for an event to become an incident requires someone there, on the spot, to observe and report. However, the regional water authorities had concentrated their pollution control staff and monitoring activities elsewhere, to deal with industrial pollution, and the other relevant professional field staff who might be alert to pollution occurrences – namely fishery biologists and water bailiffs – were thin on the ground. Such staff, nonetheless, were accustomed to cooperating closely with interested members of the public, such as anglers, naturalists and riparian owners, who were often vital sources of information on the conditions of rivers and their fish life, which included reports of pollution and fish kills. For reported incidents to be taken as a yardstick of the problem, it was necessary that potential informants be actively recruited and encouraged. Pollution control staff, therefore, began to adopt a proselytizing role as part of their efforts to establish a constituency for their work. For example, in some regions they began to give talks to local groups on farm pollution, discussing what to do if it was encountered.

It was the practice of the water authorities, and before them the river boards, to give descriptive accounts in their annual reports of the major pollution problems they had had to tackle. Typically, some statistics would be provided on the major pollution incidents that had occurred and on their causes. In Yorkshire, for example, the fourteenth annual report of the Yorkshire Ouse River Board, covering the period 1963–4, was the first to include a small section headed "Farm effluents" although no statistics were given. It was not until the report for 1966–7 of the now Yorkshire Ouse and Hull River Authority that farms appeared for the first time as a separate category of trade premises causing significant pollution incidents – 13 per cent of the total (Beck 1989: 468). In all subsequent years, annual reports featured agricultural pollution as a regular and increasingly important subject, and the practice was continued by the new water authorities formed in 1974.

Efforts were made by MAFF's Farm Waste Unit to collate these statistics, and from 1981 it began to issue annually a limited circulation of cyclostyled tables on pollution from farm wastes drawing on information obtained from the water authorities. This perpetuated the tradition whereby the agricultural community defined the problem and, in the words of the head of the Farm Waste Unit, established its "interpretation of figures whose accuracy must be suspect" (Payne 1986: 335). In 1983, however, an informal Farm Waste Group was set up under the auspices of the WAA, bringing together relevant regulatory staff from several water authorities, notably those such as the South West, Welsh and Severn Trent, who were experiencing the most severe difficulties with farm pollution. The original terms of reference for this group included the publicizing of agricultural pollution and the promotion of solutions; it came to have an increasingly influential role in wresting back from MAFF the initiative in defining the farm pollution problem, but also in raising the public profile of the issue.

In 1985, at the initiative of the WAA, steps were taken to ensure that more consistent statistics were collected from the separate water authorities and that the results received publicity. It was agreed that MAFF and the WAA should publish an annual joint survey of farm pollution incidents using an agreed set of definitions for the categorization of incidents. It was explained that "Pollution resulting from agricultural activities is of growing importance and it was decided that a regular survey would assist in determining the main reasons for pollution occurring and so indicate where improvements are necessary" (WAA/MAFF 1986: 1). The report of the survey, actually published by the WAA, was entitled *Water pollution from farm waste*.

The WAA had been provoked into an active stance in part by the prospect of the introduction of MAFF's *Code of good agricultural practice*. The WAA's Farm Waste Group coordinated the water sector's response to the draft code and this had helped greatly strengthen the preventive potential of the advice it contained. There was then keen interest to pursue the implementation of the new code, which was introduced in January 1985. Unfortunately, its introduction coincided with a downgrading and staff cuts to MAFF's Farm Waste Unit (despite the recommendations of the Royal Commission that this unit should be strengthened). This provided the WAA with both cause and opportunity to assume the leading role in the overall monitoring of farm waste pollution.

The survey of pollution incidents did provide a means to gauge what effect, if any, the code was having on advisory or farming practice. Unfortunately, incidents caused by silage effluents, for example, were revealed to have leapt by 76 per cent to more than 1000 in 1985 (WAA/MAFF 1986). Silage-making had increased greatly in the wake of milk quotas, as farmers had sought means to cut back on bought-in feedstuffs for their animals (Brassley 1996). This had often occurred without adequate provision for silage effluent storage facilities (see Ch. 3), and the poor summer weather did not help matters. MAFF's *Code of good agricultural practice* advised: "In favourable weather, the best policy is to wilt the crop". This would certainly help to reduce the pollution risk. However, advice in the farming press and from ADAS officials was often against wilting. As one ADAS scientist remarked "Any production system for silage that uses wilting pays a tremendous penalty on animal output [sic]" (Pearce 1986: 27). Water authority staff were concerned that MAFF was not prepared to condemn those who did not wilt, and this issue was seized upon because it seemed to make a mockery of the code. As Ron Thoms, Chief Scientist at Wessex Water, remarked "If there were to be a much closer involvement of MAFF in solving the farm pollution problem, the difficulties experienced could become a thing of the past" (ibid.: 27). The volume of incidents thus became something of a test for the *Code of good agricultural practice*. As the *Water pollution from farm waste* report commented, "It is apparent from this survey that, had the farmers concerned followed the recommendations in the Code, the number of pollution incidents would have almost been eliminated" (WAA/MAFF 1986: 5).

The development of incident statistics can thus be seen as part of an effort by water authority regulatory staff to get a handle on the problem of farm pollution as a basis for judging the adherence of farmers and their advisers to the *Code of good agricultural practice*. However, the availability of the statistics and the publicity they attracted also helped transform public perceptions. Not least, with its indication that the total number of reported incidents had increased by a quarter on the previous year, *Water pollution from farm waste* drew considerable attention for the first time to this particular, and seemingly growing, problem. Inevitably, the report informed and shaped the wider debate about trends in river quality.

Defining the problem

The river quality survey ranked stretches of water into four categories ranging from "good" quality to "bad". Overall, it found that "The waters of England and Wales continue to be of high quality with 90 per cent of river and canal length and 92 per cent of estuarial length assessed of Good or Fair Quality" (DOE/Welsh Office 1986a: 3). However, over a quarter of river length was assigned to a different class in 1985 than it had been in the previous survey undertaken in 1980: 12 per cent had improved its classification, whereas 14 per cent had moved down.

Of course, the question arose of who or what was to blame for this net deterioration in river quality. It should be pointed out – and there were those who did so – that there were difficulties involved in interpreting the results of the river quality survey and even more so in comparing the results from one year with those of a previous year. This did not stop various groups from providing their own gloss; indeed it encouraged them. The report of the survey contained commentaries by each of the water authorities on the underlying factors affecting changes in water quality in their areas. The main factors behind improvement in quality were identified as follows: investments in sewerage and sewage treatments works; improvements by industry to trade effluent discharges; the closure of factories and collieries; the diversion of polluting discharges to water authority treatment works; and the reduction of farm pollution through preventive campaigns in particular areas. The main factors in the deterioration in water quality were identified as: a worsening of the effluent from sewage works, often because of greater loads; increased operation of storm overflows because of overloading; more pollution from intensive farming practices; discharges from fish farms; discharges from mines; and an increase in pollution incidents, particularly from agriculture. Within this agenda, various interest groups and commentators began to point an accusatory finger at specific factors. Agriculture, it will be observed, appears both on the credit side (once) and on the debit side (twice). The publication of the compiled farm pollution incidents, when no breakdown of the incidents caused by other industries was published, also helped to focus attention on agriculture's role in water pollution.

This focus served the interests of a variety of influential groups, all of whom were given the opportunity to spell out their concerns when the House of Commons Environment Committee launched its inquiry into river pollution in February 1987. To industrialists it was a welcome development to have the spotlight of attention turned from polluting industrial practices onto farming. As the Confederation of British Industry (CBI) pointed out "The improved quality of known, direct discharges, including those from industry, has often served to emphasize the significance of diffuse and unknown sources of potential pollutants" (CBI 1987: 64). In a neat reversal of roles – indicative of the wider shifts in meaning that were occurring – the CBI expressed its concern about the increasing incidence of pollution from intensive farming practices because "Industry, which is a major consumer of water, is naturally concerned about any reduction in water quality which is likely to have an impact on the water's potential uses" (ibid.: 63). For most industrial users, occasional contamination of their water supplies from agricultural discharges posed a rather specific problem. As the CBI pointed out, "it is not usually the substances released into a river in agricultural runoff which have caused problems for industrial abstractors, but more the unpredictability of such discharges, which can lead to chemical imbalances within an industrial process" (ibid.: 64). Industrialists doubted the vigour with which a privatized water industry would pursue such problems where they did not directly threaten its abstraction points.

For the water authorities, likewise, farm pollution caused difficulties but also helped deflect attention from the mixed performance of their own sewage treatment works. The WAA argued strongly that the latter problem, in any case, was attributable to the financial constraints under which its members had to operate, whereas the "rapid growth" in farm pollution demanded new countermeasures (WAA 1987: 17). With their exacting standards of water quality for public consumption, the water authorities relied on abstraction from rural catchments for potable supplies of water. They were particularly concerned therefore at the impact on their operating costs of the possible need for advanced treatment and the commissioning of alternative supplies to counter the effect of farm pollution. The Welsh Water Authority had estimated that the capital and revenue costs incurred in one year in only two of its seven divisions to overcome farm waste problems was approximately half a million pounds – costs then passed onto water consumers (Welsh Water Authority 1984). The prospect of privatization simply posed more starkly the question of who should bear these additional costs.

To rural conservationists, a focus on farming provided yet another opportunity to challenge agricultural policy and its support for intensive methods of production. As the Council for the Protection of Rural England (CPRE) remarked, "the processes at work behind increasing pollution incidents from slurry stores and silage effluent are the same as those responsible for the increasing rate of hedgerow loss and the destruction of other important landscape features and wildlife habitats" (CPRE 1987: 362). There were also direct consequences for freshwater fauna and flora. The Nature Conservancy Council (NCC) implicated agricultural pollution in the demise of at least two species of dragonfly, as well as in the reduced populations of the less tolerant fish, insect and plant species (NCC 1987: 256–60). Between 1986 and 1990, the NCC found that at least 71 Sites of Special Scientific Interest had been damaged by farm pollution incidents, and feared the real figure could be "considerably higher" (NCC 1991: 15). Wildlife groups had long pressed for the implementation of specific provisions, in the 1974 Control of Pollution Act (§46(1–3)), to remedy or forestall pollution injurious to flora and fauna. Government proposals to revoke these unimplemented measures as part of its privatization plans triggered off a campaign by wildlife organizations to make nature conservation a central consideration in the national water quality objectives that the government had proposed should be set for the privatized water industry. As the NCC complained, "the criteria used for judging 'clean' rivers are not enough to guarantee that sensitive plant and invertebrate communities will flourish" (NCC 1987: 256). With the wildlife interest concentrated so singularly on the highest-quality rivers and with just these watercourses being the most vulnerable to agricultural runoff, conservation organizations focused on this source of pollution to highlight their campaign.

Indeed, the large majority of the 60-plus organizations that gave evidence to the House of Commons Environment Committee's inquiry into river pollution identified farm waste as a major source of river pollution. These included, *inter alia*, the DOE, the WAA, the Chemical Industries Association, the CBI, the Natural Environment Research Council, all of the regional water authorities that gave evidence, Friends of the Earth, Greenpeace, the Institution of Water and Environmental Management, the NCC, the Royal Society for the Protection of Birds, the Anglers' Cooperative Association, the Association of Chief Technical Officers, the CPRE, the Country Landowners' Association, the Countryside Commission, the Royal Society for Nature Conservation, the Water Companies Association, the Water Research Centre and the National Anglers' Council. Such a chorus of objections could hardly be ignored.

Making sense of the data

Few of the organizations that gave evidence to the House of Commons Environment Committee's inquiry had expressed themselves on the issue previously. Most had been unaware of the problem. Nevertheless, they seized on the emerging evidence of farm pollution to pursue their primary concern, which was opposition to the privatization of the regulatory functions of the water authorities. Few had any independent evidence of their own on farm pollution and, to interpret the river quality survey, most of them drew heavily on the compiled statistics for farm pollution incidents in the WAA's report *Water pollution from farm waste*. In a Memorandum to the Environment Committee, for example, the Natural Environment Research Council (NERC), in what was to become a routine juxtaposition, expressed the seemingly mutually reinforcing interpretation between them in the following terms:

> The last full river survey in 1985 showed that there had been a reduction in gross pollution over the preceding five years, but also indicated deterioration in some rivers. The latter was attributed to several factors but farm pollution was identified as a major problem. A report by the Water Authorities Association based on a survey in 1985 has documented the rising trend in farm pollution incidents arising from effluent. (NERC 1987: 78)

Careful commentators, though, conceded that both of these data sources were flawed. With the river quality survey there were problems to do with differences in methodology and data interpretation between the water authorities, for example in the discretion allowed in the assessment of biological evidence as against chemical parameters, in the option to take data from the previous two years and in the scope for differences of interpretation of the river classification system. Such problems were particularly acute in assessing the evidence from rural catchments where agricultural pollution would have been most prevalent. As Dr Mance (Regional Scientist of the Severn Trent Water Authority and representative of the WAA) remarked:

> [in] the rural catchments . . . sampling frequencies are low and therefore the security of actually making a judgement about which class the river is in, is correspondingly low and we are relying on support information in terms of biological survey data. There is inevitably an element of subjective judgement as to what that actually says in relation to the chemical results we have had. (House of Commons Environment Committee 1987: 29)

Such discrepancies raised difficulties in interpreting not only the aggregate results for England and Wales but also changes between successive surveys, as ministers and their senior civil servants emphasized. The head of DOE's Water Quality Division pointed out various problems in using the river quality survey to identify and interpret trends:

One is the classifications have changed. There are problems of consistency in the way the information is collected by authorities and if you are measuring things at two points in time, there are also problems of weather conditions, which can have quite an impact on water quality. (ibid.: 46)

To underline such misgivings, the DOE announced that it had instituted a review of river quality monitoring procedures, with the aim of standardizing the classification methods used and providing a better indication of long-term trends (DOE 1987: 9).

On the other hand, agriculture ministers and their civil servants were anxious to put "into perspective" the statistics on farm pollution incidents. The total number recorded in 1985 involved "less than one per cent of the farming community", pointed out MAFF, "and represents only one-sixth of water pollution incidents from all sources" (MAFF 1987: 116). In addition, Mr Reed (Head of MAFF's Environment and Conservation Policy Division) reasoned that "The increase in the number of pollution incidents may not be entirely due to an actual increase on the ground. Part of the increase will surely be due to the greater public awareness of pollution problems, to improved reporting by the water authorities and to public matters of that sort" (House of Commons Environment Committee 1987: 122). There was also the possibility, though, that *reported* incidents revealed only the tip of the iceberg. The South West Water Authority (SWWA) reported the following results from its Farm Pollution Campaign in which all farms in selected catchments were being visited: "There is surprising consistency in the extent of the problem from catchment to catchment. With little variation first visits reveal the same proportions of 'red', 'green' and 'blue' farms: 25 per cent – pollution occurring; 25 per cent – pollution potential; 50 per cent – satisfactory" (SWWA 1987: 307). From these findings the Authority estimated that, of the 30 000 farm holdings in Devon and Cornwall, between 5000 and 10 000 were causing pollution (ibid.: 307). The number of reported farm pollution incidents for the region in 1985, however, was only 622.

If analysis of the two separate data sources raised problems, then comparison between the two of them might be thought even more fraught. After all, the statistics had quite different bases. However, no-one raised this objection. Instead, different interests tended to use their interpretation of one to shore up their interpretation of the other. Although the vast majority of organizations did this to highlight agriculture's contribution to river pollution, a different gloss could be placed on the comparison. The National Farmers' Union (NFU) attempted to do just that. Its Memorandum to the Select Committee commented, "The number of pollution incidents reported by the water authorities has shown a steadily increasing trend over the past ten years, in spite of the overall improvement of river quality which has taken place in this period" (NFU 1987: 137). The Memorandum then went on to challenge the statistics on farm pollution incidents: "it is by no means certain that the increasing number of incidents reflects an increasing level of pollution . . . We would not accept the validity of this system of measurement by the number of recorded 'incidents' as an objective method of determining levels of pollution" (ibid.: 137). Extraneous factors, identified by the NFU as likely to influence the number of reported incidents, included shifts in public awareness, changes in staffing levels, fluctuations in rainfall from one year to another and the effect of the industrial recession reducing the amount of industrial pollution. Undoubtedly, there was some truth in these arguments. The recording of pollution must be dependent on the resources and effort put into it. Similarly, rainfall fluctuations do contribute to variations in pollution. For example, 1985 was a very wet year and, although this would have helped dilute industrial effluent flows, it would have exacerbated farm pollution problems. The NFU's line, though, was a difficult one to carry conviction, as it involved a reading of the river quality surveys at odds with the whole thrust of the Environment Committee's inquiry; and the investment of a touching faith in the soundness of that data source to support the Union's thorough-going scepticism towards the figures on farm pollution incidents.

In any case, by the time the NFU came before the Environment Committee, the tide of opinion was running too strongly against it. The DOE, the first organization to be interviewed by the Committee, had identified two main causes of the deterioration of river quality: agricultural pollution and increased loads from sewage treatment works (DOE 1987: 10). The WAA were the second to give evidence orally, a week after the DOE. By then, it would seem, a remarkable consensus on the problem had emerged. As Sir Hugh Rossi, the Chairman of the Environment Committee, commented in his opening questions to the representatives of the WAA, "We shall want to pursue with you . . . the questions of agriculture and sewerage because from the evidence we received last week these seem to be the two factors contributing the most to the pollution of our rivers at the moment" (House of Commons Environment Committee 1987: 28).

Agricultural pollution now had a formal existence as a significant problem acknowledged by a range of authoritative organizations.

Local investigations: the Torridge Report

Of course, the precise causes of farm pollution and practical solutions still remained to be decided. In this context, there was a premium on well constructed local investigations. The decline in river quality was most marked in the South West of England, where the 1985 survey revealed that almost 45 per cent of total river length was classified as being of lower quality than in 1980 and identified agricultural practices as the major cause of the decline. In this context, the South West Water Authority's detailed investigation into the environmental degradation of the River Torridge took on special significance as a case study of the causes of deterioration of one particular river system. The members of the Environment Committee were briefed on the Torridge Report, and several of the organizations that gave evidence to the Committee drew heavily on it.

The River Torridge was chosen for detailed study because there was evidence of water quality deterioration: its salmon and sea trout fishery had declined dramatically and there was considerable public interest in the river's decline. The Poet Laureate, Ted Hughes, fished there and had added his voice to complaints over declining catches. The study found clear indications that quality in most of the catchment had declined in the short term. The deterioration related in the main to determinants such as dissolved oxygen, ammonia, BOD and suspended solids, which are closely associated with prevailing land-use practices. The limited studies of invertebrates, macrophytes and algae indicated pollution-sensitive biota were no longer as abundant as they had been and, conversely, those species tolerant of organic pollution were more numerous and widespread.

Surveys of juvenile salmon and trout populations showed significant decreases in distribution and abundance. Areas where salmon were relatively numerous had become extremely restricted. Some tributaries were devoid of both salmon and trout, suggesting that the cause was not solely excessive cropping of salmon, a cause suggested previously. Such tributaries were often completely unaffected by sewage disposal, trade wastes or any human activity other than livestock farming.

Land use and farming practices in the catchment had changed considerably since the early 1950s. Reductions in the area under cereals, rough grazing and temporary grass had been matched by a large increase in intensively managed permanent grass. Livestock numbers had almost doubled and, in turn, this had led to increased use of fertilizers and the making of silage, as well as a move to winter housing, with greater problems of waste handling and disposal. It was estimated that there were 84 000 cattle and 140 000 sheep in the River Torridge catchment. In a telling but highly charged comparison, the potential pollution load from the cattle was calculated as equivalent to that from 589 000 people, about 40 times the existing population in the catchment connected to sewage treatment works (SWWA 1986).

There were contrasting interpretations, though, of the implications of the Torridge Report and the approach adopted by the SWWA to the problems revealed. The SWWA itself concluded there was:

> a considerable weight of evidence implicating modern agricultural practices as the main cause of river quality deterioration in the South West. This is a direct, unwelcome, environmental consequence of national agricultural policy which has concentrated on increasing production without adequate considerations of wider implications. (SWWA 1987: 308)

Friends of the Earth (FOE) referred to the Torridge report as "an unprecedented study of the impacts of agriculture on water quality" (FOE 1987: 175). Basing its evidence to the Environment Committee on water pollution from farm effluents almost entirely on the Torridge Report, FOE concluded that, "Agricultural wastes, particularly silage liquor and slurry, now cause widespread and worsening water pollution" (ibid.: 175). Likewise, the NCC's Senior Freshwater Ecologist referred to the evidence of reduction in salmon and salmonid stocks on the Torridge as exemplifying how less tolerant species succumb to agricultural pollution (NCC 1987: 262).

The Torridge Report, however, also conceded "the lack of irrefutable, quantitative evidence linking the problems [in the Torridge catchment] to the suggested major causes" (SWWA 1986: 7). The relative contribution of gross discharges as against problems of chronic runoff and of different types of pollutant, such as slurry, silage liquor and farm hygiene products, remained unclear. To the SWWA, this served to highlight the "inadequate data base on which environmental protection is founded" (ibid.: 7), and in its evidence to the Environment Committee it drew attention to the financial constraints that compromised its ability to undertake effective environmental monitoring, with the consequence that:

the authority will continue to be criticised for not fully understanding the problem [and] to be subject to attack by vested interests suggesting that the evidence linking agriculture, river quality and environmental deterioration is only circumstantial and not adequate to support change in agricultural practice. (SWWA 1987: 308)

As if on cue, the head of MAFF's Environment and Conservation Policy Division, when asked by the Environment Committee about the Torridge study, replied: "They have had a significant deterioration in their water quality . . . although it is not clear . . . what are the precise reasons . . . There is certain subjective evidence to the effect that agriculture is involved. But I do not think that we can pin down precisely what are the causes. It certainly is not agriculture alone." (House of Commons Environment Committee 1987: 121)

Inevitably, also, the question arose of how typical the problems of the Torridge catchment were, or as one MP put it to the delegation from the DOE, was the Torridge "the norm for relatively intensive dairy farming parts of the country" (House of Commons Environment Committee 1987: 23). The Head of the Water Directorate responded "this is a very interesting, but probably extreme, example in that it is a small, enclosed catchment with a very large load of intensive farming upon it. There are other examples but . . . not . . . ones that produce this kind of effect quite so clearly. This one tends to be quoted." (ibid.: 23). The more the Torridge study was quoted, the more it fixed the notion that dairy farm wastes were a major cause of the deterioration of rural rivers that must be tackled. This triggered off considerable debate during the Environment Committee's hearings over what were the shortcomings of the existing safeguards.

Challenging agricultural exceptionalism

The tide of opinion was clearly against the exceptional legal status for agriculture that the Control of Pollution Act 1974 (COPA) embodied. As Dr Matthews, an operations manager for the Anglian Water Authority, remarked "the agriculture industry has, in effect, been left outside the normal network of controls . . . By virtue of the "exemption" enjoyed by the agricultural industry, the aquatic environment is suffering unnecessary harm." (House of Commons Environment Committee 1987: 252).

MAFF alone reiterated the traditional justification for agriculture's exemption under COPA. It began its assessment of the impact of agricultural activities on water pollution with the words: "It is recognized that modern farming practices have a potential to cause pollution of water. However, it is equally important to acknowledge that such practices are necessary to meet consumer demands" (MAFF 1987: 116). Yet, MAFF was also keen to dispel the notion that its *Code of good agricultural practice* in any way sanctioned pollution. As the Head of the Ministry's Environment and Conservation Policy Division commented, "We have never found a case where a farmer would have committed a pollution offence if he had complied entirely with the Code's recommendations" (House of Commons Environment Committee 1987: 131). Dr Holdgate, the DOE's Chief Scientist, likewise commented that "adherence to the *Code of good agricultural practice* would substantially reduce, if not eliminate, the problem", and added cryptically "we are in the first instance looking at ways of operating within the Code more effectively than has been the situation so far" (ibid.: 22).

Ironically, what had been intended initially as a safeguard for farmers was now, in different circumstances, interpreted as an environmental safeguard. However, if the Code did indeed incorporate adequate preventive measures, why was it necessary to exempt compliant farmers from prosecution? This legal privilege would seem to serve no practical purpose and appeared unwarranted. MAFF knew of no case where a farmer had successfully used the defence, and of only one case where a farmer had even attempted it. Significantly, no-one was prepared to defend the privilege before the Environment Committee, nor to gainsay the WAA's recommendation of "strengthening the status of the *Code of good agricultural practice* so that it is used as a positive guide rather than a defence under COPA" (WAA 1987: 27). On the contrary, MAFF officials called for more prosecutions of offending farmers as a deterrent (House of Commons Environment Committee 1987: 123), and both the NFU and the Country Landowners' Association accepted the necessary contribution of prosecution (or the threat of it) to an effective strategy for combating farm pollution. Such rearguard advocacy of exemplary action against "rogue" individuals when voluntaristic arrangements were coming under strain had had parallels for the agricultural community in other environmental fields (Cox et al. 1988).

The position that MAFF sought to maintain was that voluntary cooperation, along the lines of the *Code of good agricultural practice*, would be sufficient to take care of the problem in most cases. The NFU, though, was not alert to the new and positive role being accorded the Code; its submission to the Select Committee failed even to mention it. One of the MPs commented on the omission and referred to his own experience: "I met seven dairy farmers, one of whom had a copy of the Code, two had heard of it but said that they did not have time for things like that and four of them did not know what I was talking about" (House of Commons Environment Committee 1987: 148). David Naish, then Deputy President of the NFU, responded lamely, "There is a general awareness but a very unsatisfactory amount of knowledge of the detail of the whole of the code" (ibid.: 148).

The Environment Committee concluded "*The Code of good agricultural practice* may be a good concept, but the reality does not withstand very close scrutiny" (ibid.: xxviii). About 20 separate publications formed the Code. They were not widely available and it would have cost a farmer up to £20 to acquire all of them. As the Head of MAFF's Environment and Conservation Policy Division commented "It would be wrong to envisage a situation in which the entire Code (which stands some six inches tall when all the leaflets are stacked together) is thrust into the hands of an individual farmer . . . So it is the responsibility of an individual ADAS officer to make a judgement as to which parts of the Code a farmer would need." (ibid.: 125). Crucially, therefore, this raised the question of the nature and quality of the advice directly available to farmers, but also whether farmers should be obliged to seek or act upon that advice. Here again various of the disputants referred to experience from the South West.

Devising a solution

The SWWA had adopted a particular approach to the problem, starting with a series of visits to farms in the Torridge catchment: to assess the pollution risk on each farm, to check their waste-handling facilities, and to explain to the farmers the effects of effluent runoff. The intention was not only to gather information to clarify the extent and causes of farm pollution but also to identify what should be done to improve matters. The SWAA had sought the cooperation of local ADAS and the county NFU in approaching and following up farmers. It had been agreed that farmers with problems would be referred to ADAS for advice. To encourage cooperation they would not be prosecuted, but problems would be followed up. A second visit was timed to allow the farmer adequate opportunity to approach ADAS and to check on the appropriateness of any proposed changes. A third and final visit was made to check completion of the improvements.

On the initial visits, about a quarter of livestock farms were found to be actually polluting, and a further quarter were in danger of causing pollution. By the final visit fewer than 1 per cent were found to be polluting. The results persuaded the SWWA to try the approach elsewhere within its region and it was decided to launch a farm campaign under the slogan "Pollution – together we can beat it." In doing so, it had the approval of agricultural interests defending the principle of voluntary cooperation. The Head of MAFF's Environment and Conservation Division commented: "the campaigns in the South West have achieved a considerable improvement [and] provide an example of what can be done, if the resources are available, through advice and exhortation" (ibid.: 122). Likewise, in proposing a joint approach by MAFF and the water authorities to pollution prevention, the NFU referred approvingly to the recent "concentrated campaign in the Torridge catchment by all the interested organizations [which] resulted in a very dramatic improvement in water quality for this heavily stocked agricultural catchment" (NFU 1987: 138). The Country Landowners' Association, for its part, advocated that the approach should be followed through in a series of county-by-county clean-ups, to be masterminded by ADAS.

The other water authorities also regarded the South West as something of a test case for an approach based on consultation and persuasion, but with their own reservations. As Dr Mance, representing the WAA, commented "All authorities actually carry out some educational programme as far as the farming community is concerned . . . Most authorities are looking closely at what South West are actually doing because they have taken it to an extreme because they have a particularly severe problem . . . Possibly we will follow down that route but it is very intensive on manpower and resources" (ibid.: 38). Asked if it seemed that the South West was succeeding in its approach, Dr Mance replied "the South West effort may yield benefits in the short term but the question I have in my mind is whether they have to sustain that level of effort permanently in terms of annual visits to every farm on all their catchments. That is clearly a large manpower input and we are effectively becoming an extension of the Agricultural

Advisory Service if we are not careful" (ibid.). The Environment Committee agreed and set down that it "would be concerned if water authorities ended up giving the advice by default which should rightfully be given by ADAS" (ibid.: xxviii).

It was evident that the water authorities were not prepared to rely on persuasion alone. That was what they saw as the major drawback of the *Code of good agricultural practice* – that it was "not directly enforceable and as such has had little benefit" (ibid.: 244). However, those authorities that had begun actively to prosecute farmers were not convinced that this was the best way to tackle the matter either. The North West Water Authority, for example, had suffered a marked net deterioration in river quality, over half of which was judged to be attributable to agriculture. As an official explained "We have prosecuted more people in recent years than most other water authorities for such incidents, but, to use an analogy, that is locking the stable door [after the horse has bolted], in that we can prosecute a farmer for making a polluting discharge, but the river is already polluted and the damage has been done. What we would like to see is more emphasis on prevention, and that goes back to additional controls in the agricultural industry" (ibid.: 100). There was, indeed, common ground among the water authorities over the need for preventive controls. As Dr Mance put it "Our evidence is that we cannot rely purely on the good intentions of farmers to follow the *Code of good practice* . . . We do require some compulsory preventive power." (ibid.: 236–7).

A number of the submissions from the water authorities and their staff called for legally enforceable improvement notices to allow them to tackle cases of poor waste handling and storage, and referred to an equivalent power possessed by the Health and Safety Executive. Dr Matthews of the Anglian Water Authority elaborated: "What we want . . . is for the water authorities to have powers to be able to go to a farm, to be able to say; 'We do not like that pile of manure over there' or 'That retaining wall is in a dreadful state. Will you please do something about it?'" (ibid.: 251).

The MPs were receptive to these arguments but were concerned that any such powers should not be entirely open-ended. As one MP put it "It is unusual to find that an Authority could have powers of such a Draconian type without some standards against which they can be measured as to whether they were reasonable" (ibid.: 236). The implication was that discretionary preventive powers needed an objective backstop. The Chairman of the Environment Committee developed the point – "of justice having to be seen to be done" – in a question to Dr Matthews:

If it were simply left to the wide discretion of . . . officials . . . who can vary in the severity of their attitude . . . , it might well be felt that we are introducing an arbitrary element to the extent that we were giving an individual power over a farmer . . . But not so if there are certain criteria laid down that, for example, if you have twenty cattle your catchment pit must have a capacity of so many cubic metres, that it must be lined in concrete, that it must have an outlet of a diameter of so much . . . , or whatever. Would regulations of that kind . . . be of more help to everybody concerned? (ibid.: 253)

Dr Matthews replied:

They would be of enormous help to the water industry and to the professionals working in them. It would provide a technical reference point to which all arguments could be addressed . . . If, in the absence of any technical criteria, we were to try to enter a farm to do something to use these powers, we would be involved in quite horrendous arguments, quite apart from such matters of access and magistrates' orders and so on. (ibid.: 253)

Subsequently, the Environment Committee visited a number of European countries specifically to learn how they controlled farm wastes: Members returned to Britain "convinced that it is perfectly feasible to regulate farming activity relating to slurry and silage liquor" (ibid.: xxix).

A new regulatory framework

The Commons Environment Committee concluded that rising pollution from farm effluents was an important contributory factor to declining river quality, and called for "a far more interventionist and regulatory approach to farm pollution" (ibid.: xxx). The Committee's report, issued in May 1987, recommended that the special defence from prosecution for farmers should be repealed and that a revised *Code of good agricultural practice* should be made both readily available and enforceable. As an immediate step, regulations should be introduced on the location, construction and maintenance of waste storage facilities. Grant aid should be available for farmers towards the costs involved, and ADAS should provide much more advice on pollution prevention free of charge. More generally, the Committee urged that water pollution control staff should seek to enforce the law and to prosecute polluters more frequently; and that magistrates should show less leniency.

The government's response to these recommendations followed on revised proposals for privatizing the water industry, which this time conceded the case for retaining regulatory functions in the public sector. Within MAFF and the agricultural lobby there was still resistance to the introduction of controls, which was backed up by the argument that time and encouragement were needed for farmers to improve matters, and reinforced where necessary by a tougher approach to prosecution. At the 1988 Staffordshire Silage Conference, the Minister of Agriculture warned "We either make sure that we solve these problems ourselves, voluntarily within the Code . . . or we are bound to find that the rest of the community demands that we be restricted and policed in a way which we would find burdensome and restrictive" (quoted in Beck 1989). In the foreword to the WAA/MAFF report on farm pollution incidents in 1987, which recorded yet more rises in the number of incidents, the Minister urged that "1988 must be the year when the farming community gets really tough with pollution". To reinforce the message, the Lord Chancellor agreed to exhort magistrates, when trying cases of farm pollution, to impose "fines high enough to establish that this type of illegal act can be seen not to pay" (*The Magistrate*, April 1987). The NFU circulated a letter to all its members, pointing out that farmers who polluted rivers were increasingly likely to face prosecution and urging them to take immediate preventive action. Members were advised of grant assistance and of special discounts secured by the NFU on a range of equipment to help solve some common pollution problems.

However, the WAA/MAFF report for 1988 showed a continuation of the upward trend in farm pollution incidents. These successive annual surveys maintained the profile of the issue and fed an increasingly censorious press coverage. Newspaper articles with headlines such as "Farms destroy rivers" demanded firm action from the government and tough controls (*The Sunday Times*, 5 March 1989). A "wait and see" approach was no longer credible.

The government issued its response to the Environment Committee's report in July 1988 (DOE 1988) and in November it announced the replacement of the old farm capital grant scheme with a new environmentally oriented scheme. The resultant package of measures, which was broadly along the lines recommended by the Environment Committee, tended to be characterized as a "stick and carrot" approach, with new Farm Waste Regulations and a tougher approach to prosecutions, matched with advice, grant aid and R&D. The sticks were to be under the control of the new National Rivers Authority (NRA), which was to assume responsibility for water pollution control on the privatization of the water authorities, and the carrots were MAFF's.

Under the new Farm and Conservation Grant Scheme, which came into effect in February 1989, the level of grant aid from MAFF for investment in waste storage, treatment and disposal facilities was raised from 30 to 50 per cent, and provision was made for up to £50 million to be expended on this over the first three years.[2] Every proposal for grant had to be vetted by the new NRA. Any new waste facilities were subject to Regulations, introduced in 1991 and policed by the NRA, specifying minimum technical standards such as component lifetimes, construction quality, storage capacity and siting. The NRA was also given the power to compel farmers to upgrade any existing facilities that it judged presented a significant risk of pollution. With their enforceable standards the Regulations represented a departure from the voluntary and discretionary approach that had typified British pollution control policy in general and the treatment of agriculture in particular.

The grant scheme complemented changes in MAFF's advisory policy. In 1988 ADAS underwent a major reorganization, which resulted in greater prominence for work on farm waste and pollution prevention. Since the previous year it had been obliged to charge farmers for most production-oriented advice, but "public good" advice, mainly on conservation and pollution, remained free of charge in the first instance (follow-up work, say on the design of a farm waste management plan or system, would require a fee). For some staff in the western half of the country, advice on waste management now came to dominate advisory time. For the year 1991/2, MAFF set an objective of 5000 free pollution advice visits for England and Wales, allowing roughly one in six dairy farms to be visited. The advice given drew upon and reinforced a new *Code of good agricultural practice for the protection of water*, issued by MAFF in July 1991 and made readily available to farmers (MAFF/WOAD 1991).

The new NRA was also concerned to educate and inform farmers and, although prosecution for pollution incidents was the high profile "stick", the Authority saw persuading farmers to prevent pollution as much more preferable than taking them to court. Nevertheless, the NRA did present a tougher image on prosecutions than the old water authorities had, and the ceiling on the fines it could seek to have imposed was raised from £2000 to £20 000 under the 1990 Environmental Protection Act.

Conclusions

Seldom could a field of regulation have been so dramatically transformed, involving a new regulatory regime, novel instruments, new and reorganized agencies, and reshaped policy communities. Agriculture passed from being one of the least to one of the most formally regulated sectors from the point of view of pollution control. The most profound changes were in meanings and perceptions.

A few years earlier, farm waste pollution had barely existed as a public problem; now it was recognized as "one of the two major threats to water quality" by the respected House of Commons Environment Committee (1987: xxvi). A neglected problem had thus belatedly been acknowledged. However, this was not simply a matter of public discovery of something that had been hidden; rather, it arose from a shift in wider moral purposes. Thus, the Environment Committee could ringingly proclaim: "We cannot think of circumstances where pollution of a stream or a river by a farmer could be justifiably excused on the grounds that 'it accorded with good agricultural practice'. The two are mutually exclusive" (ibid.: xxvi). Yet it was only three years earlier that Parliament had approved the coming into operation of the very section of COPA that had given farmers the "good agricultural practice" defence from prosecution for causing water pollution.

What separated the mentality of COPA from the new mentality was the altered notion of *goodness* in agriculture. In the 1970s it implied an heroic activity committed to the vital and laudable aim of providing ever more of the nation's food needs. However, with the imperative of food self-sufficiency superseded by surpluses, and with war-time shortages and rationing a fading memory, this once-hallowed priority now seemed tarnished and redundant. The successive budgetary crises of the Common Agricultural Policy had altered the public image of farmers. The once-honest sons of the soil could now be portrayed as grasping individuals demanding ever more public subsidy to intensify production in a way that damaged the environment, merely, it seemed, to add to the "mountains" of unwanted food. In the circumstances it could no longer plausibly be maintained that farm production requirements were superordinate.

In a paradoxical way, the exceptional treatment previously accorded to agriculture made it a prime candidate for subjection to the new environmental morality and its framing of the notion of goodness. There were concealed problems to be uncovered: much of the press coverage, for example, adopted an exposé stance of "the hidden menace down on the farm" kind. Critically, the discovery depended on the statistical invention of the "farm pollution incident" in a context in which agricultural practices and their consequences for the environment were coming under increasing scrutiny. Before the recognition of agricultural pollution, the debate around the Wildlife and Countryside Act 1981 had first aroused public concern. The Act formally acknowledged a problem of damage to the rural environment caused by modern agricultural practices and put in place a set of measures to address the problem. However, these new measures were based largely on voluntary means (Lowe et al. 1986, Cox et al. 1988) and so relied upon the trust and cooperation of the farming community. As claims that these measures were faltering began to take hold by the mid-1980s, then increasingly the role of farmers as guardians of the countryside was called into question. The public acceptability of systems of self-regulation rests on trust; if that trust is lost, then what once seemed proper concessions may appear to be indefensible privileges. This is exactly what happened in farm pollution policy: in the event, no-one was prepared to defend agriculture's legal exemption. As a result, the *Code of good agricultural practice* was inverted: the farmers' safeguard became an environmental safeguard.

Environmental morality was a ubiquitous feature of the late 1980s, and was not only visited on agriculture. It was particularly pronounced in regard to water, which of all the natural elements is the one most intimately associated with notions of health, vitality and purity. This partly explains the shock experienced by many at the prospect of water privatization, which seemed to shake their very faith in government. With ministers alienating this traditional and vital public service, the defence of the public interest was left to others to champion. This elicited an oppositional coalition, which included water authority regulatory staff and environmental groups unified by their espousal of environmental morality and the need for strong, independent public regulation. Also, and perhaps crucially, industry found the prospect of regulation by a private sector company to be unacceptable.

Our analysis in this chapter suggests that agricultural pollution emerged as a side issue from this other debate. Hardly any of the organizations that so roundly condemned agriculture as a significant source of pollution had expressed themselves on the issue before 1986. Most, it must be assumed, had been blithely unaware of the problem. They were alerted to it by the publication of *Water pollution from farm waste*, which provided them with ammunition in their opposition to the government's privatization proposals. One group only had considerable knowledge of the problem – that was the regulatory staff and scientists of the water authorities. It was they who defined the problem as a public issue: who compiled the river quality survey and its regional commentaries; who collated information on farm pollution incidents; and who conducted local investigations such as the Torridge study. Their efforts to raise the public profile of the issue were orchestrated through the Farm Waste Group of the WAA, and through several outlets – the separate water authorities, their professional association and the WAA – this group paraded their evidence of the problem and their proposed solutions before the Environment Committee.

In seeking to build a constituency for public regulatory functions, pollution regulators could draw on the discourse of environmental morality propagated by the environmental movement. In doing so, they needed to publicize and help politicize the existence of significant but neglected problems of pollution. Farm waste pollution exactly fitted the bill. However, in order to define the matter as a public problem, there was a need for data. Hence, farm pollution incident statistics were first collated. These seemed to indicate a large and growing problem, one that was profoundly shocking because of the desecration of the countryside it revealed by those conventionally portrayed as its guardians. Strong legislative action was demanded and it followed swiftly.

Notes

1. Subsequently, the Institution of Water Engineers and Scientists and the Institute of Water Pollution Control joined forces to become the Institution of Water and Environmental Management.
2. Subsequently, in December 1993, the grant was cut to 25 per cent, and it was abolished altogether in the Budget of 1994.

CHAPTER FIVE

The Pollution Inspectors' accounts of farm pollution

Introduction

The 1989 Water Act gave the new National Rivers Authority (NRA) significantly enhanced powers to tackle agricultural pollution, including controls for the first time over the storage of farm waste. Of equal importance, those powers were to be exercised in a greatly altered context. One aspect was the heightened public concern over water pollution, including widespread recognition of the problem of farm wastes. Another was the newly formed regulatory body itself and the way it conceived its role. The NRA deliberately sought to project an image sharply different from the old regional water authorities or the newly privatized water companies, even though the bulk of its staff came from the old authorities and most of their former colleagues now ran the new companies. In particular, it sought to align itself with the environmental movement and draw on the movement's moral authority.

An introductory leaflet issued by the NRA contrived not to mention either the old authorities or the new companies but presented itself not only as a new creation but as a novel force. Entitled *Guardians of the water environment*, the leaflet opened with the following foreword by Lord Crickhowell, a former Conservative minister and the NRA's new chairman: "September 1st, 1989, represented a turning point in the history of environmental protection in England and Wales. On that date the National Rivers Authority took up its duties. Overnight the NRA became the strongest environmental protection agency in Europe". To press home the point, the leaflet continued: "The existence of a powerful, impartial and independent organization with a clear statutory responsibility to carry out its duties transforms the way in which our water environment is guarded. It is an immense improvement on the arrangements that have existed before."˙ (NRA 1989b).

In this distancing of itself from its forebears, the NRA sought immediately to seize the high ground of environmental morality. The Authority had come "into existence at a time when there has never been more concern at the damage mankind inflicts on the natural systems on which we all depend". Foremost among its "far-reaching responsibilities" was "to control pollution and improve the quality of our country's river systems and coastal waters . . . for the sake not only of this generation – but of those to come". It would act "as a tough and effective regulator" armed not just with stronger powers but also with a publicly sanctioned moral authority: "The strength of the NRA goes beyond the authority invested in it by Parliament and the assurances of the government that it will give us all the necessary support. It taps as well the vast reservoir of public opinion, with its strong feelings about the importance of the environment" (NRA 1989b).

However, the NRA did not cover all sources of water pollution. The regulation of the major industrial sources was transferred to Her Majesty's Inspectorate of Pollution (HMIP) by the 1990 Environmental Protection Act, which made that Inspectorate responsible for integrated (or cross-media) pollution control. Likewise, discharges of trade effluents to the public sewers were regulated by the water companies. This left the companies as the most prominent dischargers to rivers to be regulated by the NRA. Although the NRA wished to distance itself from its predecessor authorities and their commercial successors, it and its staff had to reach a *modus vivendi* with their former colleagues, and the government relaxed discharge conditions on sewage works for a period, to avoid prosecution of the newly privatized companies. In any case, sewage pollution incidents were often caused by cross connections or a problem on the sewerage networks. In such cases it was not only difficult to apportion blame but there was also a defence in law for the water companies if there had been an illegal discharge by a third party. The NRA's scope to pursue its crusade against pollution was thus considerably circumscribed in relation to urban and industrial sources by the inter-institutional relations in which it was obliged to work.

Outside of urban and industrial areas, the most prevalent source of pollution was agriculture. Farm effluents did not fall within the ambit of HMIP and they were not usually discharged into the water companies' sewers. Most pollution from farm wastes occurred through direct discharge or runoff into the rivers that were the NRA's responsibility. The NRA had been given powers, moreover, that gave it an entrée into the agricultural realm through its responsibility for administering the new Farm Waste Regulations. The staff who had made an issue of farm pollution – the members of the WAA's Farm Waste Group and their colleagues – had been thrust into senior positions in the formation of the NRA. Almost inevitably, therefore, the control of farm pollution was central to the new authority's efforts to make good its claim to be the guardian of the water environment. As the press remarked, "the farming community would be one of the NRA's chief targets" (*Financial Times*, 31 October 1990).

Although the NRA's publicity and public relations proclaimed it to be a brand new organization with novel powers, a fresh outlook and no history, the reality was somewhat different. The new authority inherited not only the regional structure of the old water authorities but also most of their staff, facilities and commitments in the regulatory, fisheries and water management fields. In these fields, there was considerable continuity between the South West Water Authority and NRA South West, as there was with other regions. However, the NRA's first annual report (for 1989/90) declared its intention to take polluting events seriously, to pursue a higher rate of enforcement than the former water authorities had, and to press for stiffer penalties following successful prosecutions.

It was on the shoulders of the regional regulatory staff that the NRA's high-profile commitments vigorously to tackle pollution and defend the water environment rested in practice. This demanded a new approach of them, most especially in regard to farming. The chapter considers how that worked out in practice. After a brief description of the regional organization of farm pollution control, we recount a narrative of a day in the working life of a rural Pollution Inspector, so as to convey as vividly as possible what the work entailed. The second half of the chapter then analyzes the Pollution Inspectors' perceptions of pollution and farming, and their regulatory and prosecuting roles.

The regional organization of farm pollution control

The South West of England was one of the smaller NRA regions, having (in March 1992) 531 staff. These staff were divided between departments of environmental protection, engineering, finance and administration. The region's multifunctional Environmental Protection Department, incorporating pollution control, fisheries, water quality and water resources, had suffered from a legacy of underfunding in a region without major traditional problems of water quality, where the overwhelming issue had been seen as ensuring water supplies that are particularly stretched through the peak summer tourism period.

For the deployment of field-based staff, the region was divided into two areas. For the Eastern Area, which covered most of Devon (plus small parts of Dorset and Somerset), there were 11 Pollution Inspectors – nine with their own geographical patches and the other two helping out where needed – overseen by two more senior staff. The Pollution Inspectors were responsible for all water pollution control within their patches, including the monitoring of major dischargers. Those with rural patches tended to spend the majority of their time on farm pollution control. As the frontline of the NRA's enforcement of water pollution law among farmers, they had powers to take formal samples with a view to prosecution, to give the NRA's agreement to proposals under the Farm and Conservation Grant Scheme (F&CGS), and to issue notices requiring waste facilities to be improved. In the same Eastern Area there were 12 wardens, most of them former water bailiffs, who carried out routine work for both the Pollution Control and the Fisheries sections.

In NRA South West, strategic importance was attached to a regulatory effort which targeted the individual farm and its pollution problems, to encourage a general improvement in farmers' waste facilities and practices (see Ch. 4). This regulatory style had been developed by the former South West Water Authority (SWWA) in the early 1980s, when an ethos of "information and persuasion" prevailed. Only those farmers who persistently ignored SWWA warnings and advice had been in danger of prosecution. With the establishment of the NRA, the region presented a tougher regulatory image. As a spokeswoman declared "we are really cracking down now on farmers who pollute" (*Western Morning News*, 1 June 1990). However, work at the farm level was still regarded as crucial and, in practice, persuasion was still preferred to prosecution (see Table 5.1).

The SWWA's Farm Campaigns were continued by the NRA and were taken up in other regions, but they were no longer the main cornerstone of agricultural regulation. Following the instigation of the F&CGS in 1989, grant approval visits became an important means of contacting and regulating farmers. Pollution Inspectors visited all farms where a pollution

Table 5.1   Prosecution in the South West[1]

	Under South West Water Authority				Under NRA South West		
	1985	1986	1987	1988	1989	1990	1991
No. of prosecutions instigated	23	31	19	22	30	40	26
% of reported incidents	3.7	3.7	3.4	3.2	5.4	5.1	3.3
% of serious incidents	30	42	5	6	20	23	n/k[2]
% of major incidents						56	104
average fine[3] (£)	136	136	141	322	350	425	655
range of fines (£)	0–250	0–300	n/k	n/k	0–2000	0–2000	0–4000
total fines (£000)	**2.72**	**3.8**	**3.1**	**15.2**	**14.0**	**18.3**	**17.3**

Notes
1. Prosecution rates relate to reported incidents occurring in the stated year for which prosecution proceedings have been started. Prosecutions of incidents occurring in a previous year are not included.
2. Serious incidents were no longer reported after 1990.
3. Figures relate to cases heard in the year in question and do not include costs. Average fine levels are calculated from lists in Water Authority Association and NRA reports. Absolute discharges and cases found not guilty are excluded from the calculations. Conditional discharges are included, together with cases where no fine is actually awarded for other reasons. If more than one incident occurred at the same place on the same day and concerning the same type of pollutant they are counted as one case. If more than one incident occurs (i.e. an oil spillage and a silage effluent incident) they are counted as two separate cases.

control grant had been applied for and subsequently revisited them to check that any approved work had been properly completed.

A significant and growing proportion of the time of Pollution Inspectors was spent responding to pollution incidents. In 1990, NRA South West set up a round-the-clock pollution "hotline". In a publicity drive to advertise the free emergency number, tens of thousands of pocket-size cards were distributed around the region, along with posters for notice boards. The public were urged to carry the cards with them as a handy reminder of the number. "They can then help guard the water environment by ringing in with reports of suspected pollution" remarked the NRA's own news sheet (*The Water Guardians* no. 7: July 1990). Clem Davies, the region's Environmental Protection Manager, who had masterminded the scheme, commented "Everyone has a role to play in caring for our rivers. There is great scope for an increasing partnership between the community and the NRA . . . The Authority has staff in the field but the public are our eyes and ears too." (ibid). In the week after the launch, 50 calls were received, and more in the following weeks. Many local organizations requested additional cards to distribute, including fishing and canoe clubs, parish councils, nature groups and the coastguards. The scheme even stimulated the formation of a group called the Devon Emergency Volunteers to take on the role of a "pollution watch".

Pollution Inspectors normally received several calls a day via the hotline, varying according to the time of week (with more at weekends when more people are out and about in the countryside), the nature of the patch and the weather. Such calls were valuable in tracing farm pollution and, depending on the consequences and circumstances, could result in a prosecution. NRA guidelines distinguished the response to be adopted according to whether the pollution incident was major, significant or minor. It was national policy that major incidents should be prosecuted where there was sufficient evidence.[1] However, discretion could be applied to significant incidents[2] which might result in prosecution or a warning letter, depending on the severity of the pollution and any mitigating circumstances. Final decisions on prosecution were made by senior pollution control officers, in liaison with the Legal Department, once an incident had been identified and suitable evidence collected by the Pollution Inspector on the ground.

Reports by members of the public tended to be dominated by minor incidents, especially involving oil, as this is easily seen, even when only small amounts are involved. Large numbers of such reports could deflect the Pollution Inspectors from their other duties. However, they relished the challenge of a non-minor pollution incident, or for that matter a chronic pollution problem, which demanded painstaking detective work that tested their field knowledge in tracing and diagnosing the cause. Pollution Inspectors also kept an eye on the general state of the watercourses in their daily journeys around their patches, but systematic or intensive checks could only really be made in slack periods, such as during long dry spells in the winter. Finally, tending one's patch involved a fair amount of public relations, keeping in touch with regular informants, giving talks to farming and angling groups, and staffing a stand at agricultural shows and the local agricultural college's open days.

To convey what, in the round, was involved in patrolling a patch, we now recount a working day of a rural Pollution Inspector whom we have called Bob.[3] The account given here is a composite one, drawing on the shadowing of a number of Pollution Inspectors. Bob is therefore fictional. However, all of the component elements of his day given here – the incidents described and the views expressed – are faithfully taken from field observations. Although each Pollution Inspector shadowed was different and each working day varied, there were aspects common to most working days and to the approaches adopted by the individual Inspectors. We have presented this composite account to illustrate the range of issues and tasks Inspectors regularly dealt with and the ways in which they generally went about their work, in order to give a flavour of the job while protecting the anonymity of those we shadowed (for more details on this strategy see Revill and Seymour 1996).

A day in the life of Bob: the field culture of a rural Pollution Inspector

Bob is responsible for pollution control in the catchments of two large rivers set in a rural area expanding in population. Agriculture, however, remains the most important land use and is dominated by dairying. This is Bob's patch: a landscape of low hills scattered with farms, villages and a couple of small towns, defined by the catchments of rivers from their headwaters to the sea. Bob is responsible for controlling all sources of pollution within his patch, whether it be from the sewage treatment works of the small town – the recent performance of which has been so poor that a prosecution is likely in the near future – the large creamery processing local dairy produce, or the multitude of farms and other small rural businesses. He has a qualification in agriculture, from a national agricultural college, and experience of working on farms. Some of the recent recruits to the pollution control force have a similar background, whereas others were trained in environmental sciences.

Bob is aware that he is in the public eye. He is careful to follow up all the incidents that the public report, and to inform those who leave their names of the outcome of his investigations. He also feels it important to be seen out and about in his patch. People are then reminded about the NRA and feel it is "really doing something". They are also more likely to report any changes in their local river or stream. Conversely, his active presence locally may deter anyone tempted to discharge to the river, or may shake the complacency of those with inadequate pollution control. Even when Bob is not around, posters scattered about his patch in libraries and on parish notice boards ask the public to report pollution to the NRA. Bob himself notes several similarities between his work and that of a police officer on the beat.

Bob has an obvious pride in knowing his patch. "There's no substitute for knowing every bridge and every stream", he says. Such knowledge takes years to build up: now he rarely, if ever, refers to a map. Subconsciously he plans his routes so that he crosses the rivers and streams at various points and can keep an eye on them. Bob has a good idea of the condition of his rivers. He distinguishes particularly bad tributaries from those that are very good. He is also well acquainted with the local topography and farming pattern. He knows where small farms predominate and where the larger units are located. Dairy farms cause him the most problems; he has "far less trouble" from beef and sheep production, and even intensive pig and poultry units are less of a problem than dairy farming, where the cattle are in and out all the time. However, because he has so many responsibilities and a patch of substantial size, he cannot know the majority of farmers individually, as he does the staff at his handful of industrial dischargers. Instead, relationships with farmers tend to be formed where a "problem" is discovered or an improvement planned, and to be of an intermittent and limited duration. Indeed, there are farms in his patch that he has never visited. Nonetheless, Bob still values a personal approach with farmers. At the beginning of his work on the farm campaign visits, he went through the farmers in the Yellow Pages and reorganized the list according to farm name. Being able to go onto a farm and say "Hello, Mr X" is, Bob feels, "worth its weight in gold".

Bob has followed the encouragement of Pollution Control and lives in his patch and works from home. He works alone, although he liaises closely with the wardens who cover the same area. Bob spends as much as three-quarters of his time driving around attending appointments and emergency calls. He can easily clock up 2000 miles in a month.

He begins his day by phoning the office for information about incidents, messages from Pollution Control or any extra duties he might have to take on. He plans his day around the incident work and any appointments he has made, fitting other routine duties into any gaps left. Planning itself has to be flexible, as urgent incidents take priority. He remains in touch with Pollution Control via his car phone, and his day is usually punctuated by at least a couple of incident reports from members of the public. Sometimes he feels frustrated that, with so many incidents – usually of a minor nature – coming in, he is left with less time for preventive work, although he would not want the public to stop reporting.

His car is packed with any equipment and paperwork he might need during the day: bottles for routine samples, the formal sample kit, a camera for collecting photographic evidence, ropes, buckets, funnels and any other contraption he has, by experience, found to be of use in his daily work. Forms for the F&CGS and for making planning applications, information sheets, names and addresses and publications (such as MAFF's *Code of good agricultural practice*) are also packed in, together with the essential wellington boots and waterproofs. His car is really a mobile office and store in one!

Today, the winter weather is cold but dry, and Bob thinks he is unlikely to be inundated with incidents. The only incident-related work Pollution Control has passed onto him is to check on a sewer that had been overflowing. He has also made an appointment for a F&CGS visit and he intends to do some routine sampling at the sewage treatment works and in one of his troublesome tributaries.

On the way to the F&CGS visit, Bob diverts slightly to pass a "problem" farm, classified as "Red", or polluting, by the region's farm campaign. Like many Devon farms, this one is somewhat off the beaten track and a stream runs close to the farmyard. Bob makes a quick and discreet check on the watercourse (which in the past has had effluent going straight into it from the yard), but he does not enter the farmyard to see the farmer today. Bob often visits here both to monitor the pollution and to stress to this rather resistant farmer that the problem will not just go away. The farmer is not keen on the attention Bob is paying him, but tolerates it. Today the ditch seems in order, although it is difficult to tell when it is so dry. Even though Bob suspects that the yard was sometimes scraped into the stream (the farmer escaped a prosecution "by the skin of his teeth" in the last year of the regional water authority), his preference remains to improve the situation and avoid bad feeling with the farmer. He feels that taking a farmer to court risks destroying the relationship that has built up between them, which he regards as his main instrument in improving water quality.

Bob is only infrequently involved in court cases and does not often take formal samples, but the basic procedure is one of his field skills and he can cite the legislation, chapter and verse.

Bob arrives at the farm, a low-lying property bordered by a substantial stream, for his 11 a.m. appointment for the F&CGS approval. He already has some idea of the scheme the farmer, Mr Shields, wishes to undertake and has looked out information on past dealings with the farm. It was visited a few years ago by one of the wardens, who found slurry overflowing from an open pit. The farmer has since installed a steel slurry store to deal with that problem. However, last spring when Bob was checking the nearby stream, he found some bad discharges, and visited the farm again. He had made a note in his diary to return the following winter to check on the situation. However, before he could do this, the farmer had phoned into the NRA office, requesting a grant approval. Bob welcomes any such initiative on the part of a farmer; it is better than him having to chase them up.

On this occasion, Bob parks in the yard and goes to find the farmer. He introduces himself as "Bob Brown – National Rivers Authority", and addresses the farmer by name. Mr Shields unfortunately has lost the form he is supposed to have filled in for the NRA, but Bob has a spare blank copy. The only form Mr Shields produces is one relating to discharge consents (see Ch. 3). He asks Bob if it is worthwhile keeping the consent. Bob says no, firmly: consents are not suitable for farm effluents, which are far too polluting to comply with normal consent conditions. It is important to prevent any discharge, Bob argues, and, after expressing some initial doubts, Mr Shields concurs. The farmer does have contractors' estimates for the proposed scheme – a dirty water disposal system, including a settlement tank and low-rate irrigation, to add to the steel slurry tank he already has – and shows these to Bob. The estimates include a couple from leading engineering firms in the area, but Mr Shields would prefer to use his local builders as they are cheaper and, because they are looking for work, can start the job as soon as he gets MAFF approval. Bob peruses the estimates briefly.

He then asks the farmer to spell out what he is intending to do and to point out where the new system will go. The farmer responds with a half-hour tour of the farmyard, in which he answers Bob's enquiries. Bob wants to know what is kept where. Where are the cattle housed, where is the collecting yard, dairy, parlour and silage clamp? He places great importance on the farmer showing him round and describing the new system.

Bob is particularly concerned about the location of underground pipes and old drains. Will the existing effluent drainage, including pipes from the silage clamp, be connected to the new system? Bob asks to look at the actual site of the proposed settlement tank to check on its suitability. He then turns his attention to roof water and guttering. Is rain water diverted away from the dirty water? Bob stresses that the farmer should do as much of this as possible, as it cuts down on the volume of waste produced.

Mr Shields obviously sees Bob's visit as a good opportunity to clarify the NRA view on some issues. Although Bob is willing and able to oblige on certain points, he avoids being drawn into others. When Mr Shields seems worried that his slurry store may become solid if most of the liquid is diverted to the proposed low-rate irrigation system, Bob suggests that the new system should be flexible enough to allow the diversion of some water into the slurry store if and when this is required. On the other hand, when Mr Shields raises a point which he has discussed with the contractors about whether or not he will actually need to put in a settlement tank, Bob is less forthcoming. He warns about the risk of settlement within the irrigation pipes and thinks that much depends on the type of pump chosen. Some will finely chop the waste, thus reducing the risk of settlement within the pipes. However, he admits that he is not an expert and cannot advise on the type of pump to be installed.

After looking around the farmyard, Bob asks the farmer to show him the fields proposed for irrigation. They view the fields together and Bob, applying his knowledge of the local soils, is happy with the amount of land available for spreading. However, two of the fields are low-lying and adjoin the stream, and Bob warns these should be used for irrigation only during the summer months.

He fills in the details of Mr Shields's proposed scheme onto the F&CGS application form and the farmer signs it. Bob says the NRA's letter of approval will be sent through as soon as it is typed up, which takes at least a week. He advises Mr Shields to get on with the work as quickly as possible, as the grant scheme could run out. He then goes back around the farmyard and in a few minutes completes the sketch that the farmer should have supplied on the form. It includes the position of the dairy, cattle housing, steel slurry storage tank and reception pit, silage clamp, collecting yards, parlour, main underground drains and the points for the irrigation hydrants.

On leaving the farm, Bob pulls into the side of the road to draft his letter of approval. This is based on a standard format used in the South West region and has a series of get-out clauses for the Authority in the event of an "approved" system polluting. Bob also makes a note in his diary to revisit the farm in about three months' time to check on the progress of the scheme.

Once Bob has finished this paperwork, he carries on up the valley to make a routine check on the state of the river. He takes a route that frequently passes over the small bridges crossing one of his more troublesome tributaries, the Yore. Driving over one such bridge, Bob notices a discoloration in the stream and pulls over. He assesses the conditions by looking for visual changes in the stream. Is it cloudy, foaming, or full of sewage fungus? He smells the water. Does it smell of slurry or domestic sewage or silage liquor? Checking involves him walking right into the stream. From there it is easier to spot small amounts of the slimy white fungus that is a sign of sewage pollution. He can also pick up and feel the stones on the river bed to discover if they are slimy. Another of the clues Bob uses to indicate the presence of organic pollution is the invertebrate mix of the watercourse. To check this, he picks up stones from the stream bed and examines the wildlife on their undersides. Mayfly larvae, together with freshwater snails, are the main indicators of clean water at this time of the year. This stream, despite the discoloration, still has plenty of mayfly larvae. Further investigation of the locality jogs Bob's memory – there have been problems here before – and he recalls that an informant lives nearby. He has been out to investigate her calls a couple of times in the past, but, by the time he has arrived, the problem has disappeared. He decides to call in, both to ask about the stream and because such visits are good for public relations. She allows Bob access to the stream from her garden. While he examines the water, she tells him of a more recent incident when she found a few small trout on the riverbank, and suggests the cause was slurry runoff from local farms, although she also complains about sewage leaking from the village council houses. Bob finds information from such informants useful and well meaning, if not always completely accurate.

Bob remains unhappy with the state of the stream and wants to investigate it further, but back inside his car, he receives a call from Pollution Control informing him of a reported oil incident. He sets off immediately to deal with this, as it sounds relatively serious. This entails driving to a large village on the lower reaches of his other main river. Once he has finished dealing with the oil incident, Bob finds a quiet lay-by to have his lunch.

After lunch, Bob decides to carry on with his investigation of the Yore tributary. On his way back, he stops to check the stream below a large tenanted farm. When he was last there, the previous year, dirty water from the cattle yard was running down through rough ground and into the stream. Now the stream is very clean and Bob feels satisfied. It is one of the rewards of the job to see a once dirty stream clean and he decides to call in at the farm to see how the problem has been solved. Mind you, he is not surprised to see the improvement here. Mr Lorton is a resourceful farmer running a large farm, the type who is conscious he will have to do something about pollution sometime, but who needs a visit from Bob to nudge him into action.

Mr Lorton is very pleased to show Bob what he has done, confiding, "We should have done it years ago". The farmer has designed the system himself. All the waste scraped from the yards, the dirty water and some of the parlour washings now go into a new lagoon. The liquid fraction is pumped out and irrigated using a tanker. Although his pollution problem has apparently been solved, Mr Lorton still wants to make some improvements to his system. When the cows go out in the spring, he aims to concrete the area next to the lagoon to make it easier to scrape. He also plans to divide the lagoon with ex-railway sleepers to retain the more solid material at one end by a type of "weeping wall" effect. He hopes this will avoid the pump blocking. Bob suggests that he might consider extending to beyond the sleepers the pipe bringing in the dairy and parlour washings, to aid solid and liquid separation. Mr Lorton has not lined the slurry lagoon and laughingly notes to Bob that there must be some percolation. He and Bob consider the likelihood of this material eventually finding its way down the hill to the stream. In Bob's experience such lagoons generally seal themselves, but only time will tell, the real test coming with prolonged winter rainfall. Mr Lorton does not want problems with the stream but also hopes the lagoon will not seal itself too effectively, as seepage is a good means of disposal and it reduces his need to pump the lagoon out as frequently.

After this short stop, Bob sets off for the Yore tributary once again. Moving upstream from his point of investigation in the morning, he observes, from another of the small bridges, that the water looks clearer. However, on checking the stones he feels there is something wrong with the invertebrate mix. A farm up the valley has a small creamery with consent to discharge the effluent from its treatment works into the stream. There have been serious problems here in the past and, although thousands of pounds have been spent to remedy them, he suspects there is still something wrong. His view is confirmed when he sees the slightly discoloured and lifeless condition of the stream immediately below the discharge. Instead of mayfly larvae, Bob finds plenty of leeches and only a few small snails. He takes two samples around the discharge point and one upstream. Bob walks downstream of the creamery to where the drainage of another farm enters the Yore. On this occasion he is pleased to find that the water is fine, despite problems there the previous winter.

Bob decides then to check a little farther down stream. Although the farm campaign was carried out here three or four years previously, Bob's experience is that problems do recur. Farms need monitoring, particularly to ensure that proper maintenance is being carried out. Bob consults his map and his farm list from the Yellow Pages, and finds the farm he wants to check. It adjoins the stream and has had pollution problems in the past. Checking farms like this, where there is no nearby bridge, is difficult, as you must either enter the farmyard itself – something that cannot be done every fortnight or the farmer would begin to feel harassed – or trek up the river. On this occasion, he decides on the latter course and, after negotiating a gate and a fence and crossing a riverside meadow, Bob reaches the boundary of the farm. He thinks everything looks in order today. The ditch connecting the farmyard to the stream is clean, whereas it had been full of slurry last time. Bob is once again pleasantly surprised, and begins the walk back to the car, along the opposite bank. However, soon after he has set off, he spots a pipe which had been hidden from view from the opposite bank. Bob enters the stream to inspect the pipe more closely. Although the slight discharge from the pipe is clear, when Bob looks more carefully and feels inside the pipe he finds it is coated with a thin film of white slime. This he associates immediately with past slurry discharges. He suspects these will recur, and makes a mental note to call back when the weather breaks.

Back at the car, another call comes through from Pollution Control. A complaint has been received about a polluted pond. Supplied with the details, Bob sets off to investigate. The informant has just moved to the area and has found dirty smelly water in her pond. She introduces Bob to another nearby resident who confirms that there have been persistent problems with both the pond and the ditch. Bob discovers fungus in the pond, which he thinks is probably caused by farm effluents, especially when he finds the ditch feeding into the pond to be full of fungus. The ditch is fed by a drain and Bob sets out to trace the source of the effluent. His suspicions fall upon a nearby dairy farm. Bob parks in a lay-by and walks down the road from the farm, following the line of drains from the farmyard which sound as if they are carrying about the same amount of water as issued from the pipe into the ditch. Eventually he discovers that the drains do cross the road and are piped under the field, emerging into the polluted ditch and pond.

Having thus confirmed the source, Bob returns to the farmyard, where he meets Mrs Chapman who runs the farm with her husband. He introduces himself and tells her about the polluted ditch and pond. Her first reaction is to say "if we have to do too much we'll have to get out". This is a small enterprise with only 32 dairy cows, little land around the farmstead and no chance of buying more on the edge of an expanding commuter village. Bob is not put off: in his view, if a farm is polluting, some improvement must be made, however small the enterprise. Nevertheless, Bob cannot tell what might need to be done until he has been shown around, which Mrs Chapman somewhat reluctantly does. Bob is particularly interested to find out where the drains go and is concerned when he is told that all the dirty water is being filtered to one small field. He tells her he would like to see a broader dispersal. He also suggests putting up more guttering to divert clean roof water. Once he has looked around, Bob feels certain that the dairy and parlour washings go down the roadside drains, instead of into the field, as he had suspected. He deduces this from the straight line of drain covers that run from the dairy and parlour complex to the road drains. Bob tells Mrs Chapman that she needs to call in a consultant and he shows her a list. She is under the impression that ADAS will charge for a visit, but Bob puts her right. He leaves his card and pointedly says he will call back to check on progress.

On returning to his car, Bob makes some notes on the incident. For each report he receives, a form has to be filled out in triplicate: one for the computer, one for the filing system and one for himself. Bob then heads off home; the rest of the paperwork will have to wait until the morning.

Perceptions of pollution and farming

We observed several Pollution Inspectors in their work. Their assessments of pollution were framed in absolutes. To them, pollution was a "dirty" stream or river. Seeing a once dirty stream made clean was the ultimate goal and in principle nothing less could be condoned. Such an appreciation derived from their biological training, which had impressed upon them the complex interconnectedness of ecological systems, and was reinforced by the gathering evidence of discernible changes in aquatic ecosystems, even from very low levels of farm waste contamination. Seeing the water environment as a realm of consequences underpinned a moral imperative in the Pollution Inspectors' responses to farm pollution. Although they might concentrate on the most pressing problems, they regarded pollution on any scale as wrong and they tried to avoid falling into the trap of saying a little is acceptable. One Pollution Inspector commented that "if a farm is polluting, it's polluting . . . whether the problem is caused by ten or a hundred cows, something must be done to remedy this".

In practice, Pollution Inspectors needed more workable definitions and these were very much shaped by their practical experience. There was a strong ethos among Pollution Inspectors of "knowing" their rivers, and their definitions of pollution drew upon this very intimate knowledge and inspection of the aquatic environment. A dirty river or stream was thus assessed by: visual changes – cloudy, foaming or discoloured water, the growth of sewage fungus; the presence of odour – the smell of sewage fungus or silage effluent; the feel of stones in the stream – the presence of slime; and the nature of the macro-invertebrate population of the stream – mayfly larvae and freshwater snails indicating clean water (during winter months), whereas their absence and the presence of leeches indicated to the Pollution Inspector that the water was suffering from organic pollution. Pollution Inspectors' interpretations of pollution also tended to be influenced by the types of problems they encountered. Thus, they distinguished the most serious pollution by such criteria as fish being killed, a water intake closed, several hundred yards of the river affected, or someone else's use of the water disrupted. In contrast, working assessments of pollution took little account of the hidden problems of groundwater contamination or more diffuse pollution such as that from nitrates or pesticides.

The response of Pollution Inspectors to farm pollution was also influenced by their perceptions of farmers as regulatory subjects. Startlingly, dairy farmers were viewed overwhelmingly as a group of potential lawbreakers. This perception appeared to have two roots. First, the Pollution Inspectors were acutely aware of the sheer potency and quantity of the wastes that dairy farmers had to dispose of. Secondly, farmers had long been regarded as an intrinsically problematic group to regulate (Hawkins 1984,

Watchman et al. 1988). Compared with most of the industrial and commercial dischargers with which Pollution Inspectors had to deal, farmers were seen to lack professionalism in their approach to waste management matters and regulatory procedures and standards: they usually lacked relevant specialist training, they were often ill informed of the law, and they could be obstreperous and uncooperative in their dealings with officials. Dairy farmers also had a poor reputation as technicians, in managing and maintaining their equipment, and this was seen as a major factor in pollution incidents. All in all, this added up to an impression among the Pollution Inspectors of farmers as a difficult, recalcitrant and even deviant group.

Here again, in practice, Pollution Inspectors needed more workable definitions to deal with the numbers and diversity of farmers they encountered. Although Pollution Inspectors considered any dairy farmer capable of pollution at some time, they tended to distinguish between the majority of farmers, who seemed open to encouragement to clean up their waste management, and the minority, who seemed quite uncooperative.

With many cases to deal with, the Pollution Inspectors were often obliged to make a provisional judgement of a farmer, on their initial encounter, as likely to be a "problem" farmer or a "persuadable" farmer, drawing on their past experience with other farmers. The sort of characteristics expressed to us as informing this categorization are shown in Table 5.2. They combine assessments of the resourcefulness and flexibility of the farmer (size of the farm, age and apparent alertness of the individual), with judgements of their moral worth (tidiness of the farm, disposition of the individual and evidence of guile) and impressions of their openness and responsiveness. Pollution Inspectors then calibrated their responses accordingly. Provided that major pollution was not involved, with those farmers judged persuadable the Pollution Inspectors did not adopt an overtly coercive stance, but, often with considerable patience, they would seek to persuade them to improve their facilities and procedures for waste management. With those thought to be problem farmers, in contrast, Pollution Inspectors were inclined from the start to take a tougher line.

The regulation of farm pollution

Pollution Inspectors such as Bob were the regulators of farm pollution. However, most of their work was not the formal enforcement of pollution

The regulation of farm pollution

Table 5.2 Pollution Inspectors' perceptions of farmers

"Persuadable" farmers	"Problem" farmers
Physical factors	
Tidy farmyards	Untidy farmyards
Larger farms	Smaller farms
Social factors	
Have got it up here; (nous)	Ignorant
Younger farmers	Older farmers
Attitudinal factors	
Just waiting to be told to do something	Do not like being told what to do
Willing to talk; friendly	Tell you to get lost and hope that by doing so the problem will just go away
Straightforward	Shifty; have something to hide
Will ask for information	Never contact anyone
Resigned to having to do something	Resist having to do anything

law. Instead, as we have seen, it was an amalgam of inspection, detection, investigation and persuasion. Pollution Inspectors continuously drew upon and replenished their intimate knowledge of the local river systems, against which they assessed and assimilated the pieces of information they received – from Pollution Control, from informants, from farmers – about actual or potential sources of pollution. On the basis of that knowledge and information, they expressed their judgement about the risks from particular sources and sanctioned specific courses of action. Most of the time they were dispensing not the law but their own judgements, and this in a manner calculated to enrol farmers into their way of seeing pollution. Farmers were thus alerted to the vulnerability of watercourses, the polluting potential of farm effluents, the inescapable hydrodynamics of their own fields, and the gravity of the problem of farm pollution.

105

The way in which this was done was thought to be crucial in winning over the farmer. In the absence of major pollution, Inspectors adopted an informal approach. A farmer might be causing some pollution or be at risk of doing so, but the Pollution Inspector usually sought to be "reasonable". Thus, the law was not enforced to the letter. Instead, farmers were told what the problem was and were given time and encouragement to make changes. This was seen as the best way of securing an improvement in water quality. In the negotiations into which Pollution Inspectors were thereby drawn, their stance was interwoven with nuances of friendliness, didacticism and authority.

Right from the start they liked to give the impression of personableness. They placed store on being able to go onto a farm and greet the farmer by name. They also left their cards to help the farmer remember who called and to encourage future contacts. Pollution Inspectors were anxious to avoid seeming to be overbearing in their dealings with farmers, who were not thought to respond well to being ordered around. This might "put the farmer's back up", whereas a discussion about the problem they faced was seen as more likely to encourage positive steps and a change of outlook. One Pollution Inspector referred to his handling of a situation where manure was being stored on a concrete pad right next to a stream: he "*could* have gone in heavy handed" and demanded the farmer stop using the store. However, he thought that a chat with the farmer would do the trick. Despite having only ten cows, the farmer already had a dirty water irrigation system installed, and the Pollution Inspector interpreted this previous investment as an indication that the farmer was amenable.

Pollution Inspectors recognized F&CGS visits as particularly valuable in building up cooperative relations. The farmer would already be committed to taking some action and spending money on pollution control, and also knew that NRA agreement was needed before any grant aid could be received. On these occasions, Pollution Inspectors usually felt welcomed, and they encountered much less suspicion and foreboding than on their other farm calls. They were keen to capitalize on this different atmosphere to cement their relationships with cooperative farmers.

Not only did Pollution Inspectors find that friendly relations helped in persuading farmers to address current problems, they were also believed to encourage farmers to be more open in approaching the NRA for advice subsequently. From a pollution control viewpoint it was obviously preferable for farmers to make contact when they had a problem, rather than for the NRA to have to discover and locate the pollution, and such responsible behaviour, in all but the most severe circumstances, was rewarded with relief from prosecution. Thus, when a farmer with concerns about seepage from the concrete base of his new slurry store bantered that if there was any problem "I'll be on to you!", the Pollution Inspector involved replied "That's OK with me". This reaction illustrates well the subtlety of the Pollution Inspector's role. Although the farmer might have had a different interpretation of this interchange (indeed, we analyze this exchange from the farmer's point of view in Ch. 6), and was probably seeking to spread the responsibility for any future problems with the facility, the Inspector either ignored this aspect or deliberately turned the comment around to create a positive regulatory point. He thereby avoided a possibly fraught and formal exchange over correct legal responsibilities, and sought instead to build up a trusting relationship with this farmer.

Displays of authority by Pollution Inspectors were subtle rather than totally absent. Even on an initial routine visit, a Pollution Inspector who introduced himself as "Bob Brown – National Rivers Authority" reminded the farmer that he had the authority of the organization behind him. Thus, the initial presentation was also that of a figure of authority who could enforce the law.

In general, though, the formal sanctions of the law remained implicit. Pollution Inspectors, for example, tended not to mention the potential fines in personal negotiations with farmers. This might have been unnecessary, as most farmers already had an exaggerated perception of the risks and penalties of being prosecuted for farm pollution (see Ch. 6). Conversely, Pollution Inspectors were reluctant to refer to the actual likely levels of fine (see Table 5.1 on p. 92), which they regarded as being of questionable deterrence: "Being told you could be fined £500 if you pollute because you have not spent £10 000 on a pollution control system is hardly a compelling argument".

In negotiating improvements, Pollution Inspectors felt they had to be persistent. A not untypical example of the timescale involved is provided by a farm found to be polluting a stream on the first visit of the farm campaign in February 1988. The farmer was advised to contact either ADAS or a consultant. Although the farmer received advice from ADAS at the second farm campaign visit, not made until over a year later in July 1989, the farm was still polluting. The farmer was once again encouraged to seek advice and remedy the problem. Another visit followed just four months later and this time, although the farm again was found to be polluting, the farmer was in communication with local contractors. Once the farmer had put forward a scheme, it was agreed to by the Pollution Inspector under the F&CGS arrangements. The new facilities were completed in June 1990 and, when checked in November of that year, were thought to have solved the problem in the stream. This farmer was not considered to be deviant; indeed, the Pollution Inspector involved described him as friendly and prepared to invest in a comprehensive scheme. However, the farm did have serious locational problems, which meant that the solution to the problem was more complex than normal, and between each of the five visits he had taken steps – contacting ADAS, calling in contractors, putting forward a scheme under F&CGS – which could be taken as indicating a cooperative stance. Nevertheless, it was nearly two-and-a-half years between the initial classification of the farm as polluting and the completion of the new scheme.

For most farmers the problem was not so much one of causing pollution as being detected doing so. In that sense, the Pollution Inspector was the embodiment of the problem for the farmer, and Pollution Inspectors actually saw their own persistence as a means of convincing farmers that a pollution problem would not simply go away. Individual Pollution Inspectors had ways of reinforcing this message verbally. One played upon the "peace of mind" that an agreed improvement would bring. He reassured farmers that, once they had addressed their particular problem, "at least you know we won't be bothering you again" – the apparent solicitousness laced with the threat that failure to act would attract not only continued regulatory surveillance but also possible penalties.

As the example above shows, a series of visits might be necessary to encourage or cajole a farmer into making some improvement. Revisiting also allowed Pollution Inspectors to update farmers with current NRA thinking. As one Pollution Inspector noted, practices "thought OK four years ago we wouldn't say were OK now". This upgrading was a continuing process. For example, none of the Pollution Inspectors shadowed in 1991 required farmers to plan for eventualities such as frozen or waterlogged ground in the construction of low-rate irrigation systems. However, the following year a requirement for contingency planning was laid down in national guidelines for farmers applying for grant aid on low-rate irrigation systems without four months storage. Given the poor reputation of farmers for looking after their equipment, revisiting was also viewed as essential to check that they were managing and maintaining their pollution control facilities and practices adequately. However, pressure of other commitments limited the scope for such "after treatment" revisits, which tended to be concentrated on catchments with particularly intractable problems.

Inevitably, in these repeated encounters with farmers, Pollution Inspectors conveyed a good deal of information and advice. However, they saw themselves primarily as pollution regulators and protectors of the water environment and were reluctant to admit an advisory role regarding the remedial or preventive measures a farmer should take, out of a concern not to compromise their regulatory duties. However, fieldwork identified important types of information, which they integrated into their enforcement work and which conveyed something of their own understanding of pollution. The intention seemed to be to get farmers to make connections between their own activities and the consequences for the water environment and, from this altered perspective, to reassess their responsibilities.

First, Pollution Inspectors made use of their water industry experience and expertise, supplemented by some knowledge of farming, to encourage farmers to look at their farms from the viewpoint of water protection. They would point out any problem they felt the farmer had. This might involve drawing attention to "something simple, something the farmer has never thought about". Much of this type of advice was aimed to make farmers more conscious of the hydrological characteristics of their own farms: getting them to think carefully about the location of ditches, streams and underdrains, about flood risks and about variations in slope and soil permeability. Such advice potentially had an important educational role in improving the farmer's awareness of environmental risks in their managerial decisions.

Pollution Inspectors also played a crucial role in educating farmers about the *impact* of farm pollution and more specifically of their own farming activities on the water environment. In part this constituted an appeal to the polluter's sense of social responsibility. Although the Pollution Inspectors acknowledged progress in this area – most farmers had come to regard the disposal of slurry into rivers as unacceptable (see Ch. 6) – they still came across ignorance or indifference about the effects of less publicized farm pollutants. For example, they found that some farmers, apparently not realizing that milk was highly polluting, might be controlling their slurry and yard washings, but overlooking those from the dairy and the parlour. Other farmers were found to turn a blind eye to small discharges of farm wastes, perhaps believing that, as not much was escaping, it would not do any harm. Pollution Inspectors sought to convince farmers of the damage that was caused by such small discharges, by showing them the resultant sewage fungus and its effects.

Such educative efforts were distinct from technical prescription. Not only did Pollution Inspectors eschew a formal advisory role, they also denied having the relevant technical expertise in agriculture and farm waste management. Nevertheless, they did have strong views on technical solutions to farm pollution problems, which informed their regulatory work. In practice, they often gave out information on general technical matters which might influence the type of facility a farmer decided to invest in. Thus, on farm visits they all repeatedly emphasized the importance of the separation of clean and dirty water in preventing pollution, which was "half the battle" in one Inspector's view. Likewise, Pollution Inspectors occasionally expressed their judgement on the choice of equipment. For example, the use on hilly ground of mobile irrigators, designed to operate on flat areas, was questioned. Conversely, some Pollution Inspectors explicitly favoured "weeping wall" stores, arguing that they presented less risk of pollution if they failed than an above-ground metal store would because it would be containing more liquid slurry. Some, indeed, dissuaded farmers from installing tin tanks because they believed the farmers would not carry out the routine maintenance required to minimize structural failure. In certain contexts, the need for specific advice-giving was acknowledged: for example, in the case of emergencies where stop-gap measures needed to be taken to minimize the pollution; and on "problem farms" where a solution was proving difficult for the farmer and consultant, and the Pollution Inspector was invited along to give his or her views.

Pollution Inspectors were at their most specifically prescriptive in impressing on farmers the need to manage and maintain farm waste facilities properly, usually when inspecting new facilities installed under F&CGS. At that point, they could not be accused of interfering with the main technical choices to be made, but might be less than happy with the outcome because they felt that some systems demanded more attention than farmers were prepared to devote to them. In any case, with their perception of farmers as poor managers of equipment and machinery, Pollution Inspectors were at pains to urge farmers to make daily checks on their waste storage facilities and to check nearby watercourses. They also stressed that mobile irrigators must be moved regularly, that tin tanks should be periodically cleaned out and inspected internally to minimize the risk of structural failure, and that sluices and valves should be checked regularly to guard against blockage.

In the main, where specific technical advice was needed, Pollution Inspectors tended to follow the official line of the NRA and deny that they did or could advise farmers. As one Pollution Inspector remarked, "It's not our job to advise on specific work; we're not qualified. . . . We're not consultants". It would put them "in an impossible situation" if advice were given and subsequently there was pollution, as this could jeopardize court action and harm the reputation of the NRA.

Instead, farmers were referred to ADAS or other agricultural consultants for such technical advice. Although Pollution Inspectors thus nominally deferred to agricultural expertise, in private discussion some were critical of what was available. One Pollution Inspector felt ADAS free advice to be "no better or more extensive than the advice we give". In his view, it just told the farmer he or she had a problem, and the real point of the free visit was to sell ADAS commercial services. Quite apart from such muted questioning of the exclusiveness of ADAS's authority, the Pollution Inspectors as a group saw themselves as having an important role to play, alongside ADAS, in educating farmers.

Thus, the Pollution Inspectors had to perform the sometimes difficult task of balancing the need to provide information and yet to avoid giving specific advice; as one Pollution Inspector noted, there was "always a thin line" between telling farmers they had to do something and telling them what to do. The NRA stance against giving out technical advice contrasted markedly with that adopted by field staff of the regional water authorities in the mid-1970s. Hawkins reported the willingness of those officers, in dealing with industrial dischargers, to act as "expert consultants" and to give "technical advice" as part of the compliance process in which advice was traded for cooperation. Although this advice was given out "tentatively", to protect the officers and their authorities from any liability repercussions, the lack of prosecutions minimized that possibility (Hawkins 1984: 125). With the NRA's greater emphasis on prosecution, there was now a cautiousness about being seen to give advice. On the other hand, particularly with regard to farmers, there was also much greater opportunity and pressure to advise.

The threat of prosecution

The crucial difference with the past was that prosecution was now re-garded as an active option, reinforced by internal and external pressures on the NRA to take a firm stance against pollution and its perpetrators. In this context, Pollution Inspectors saw themselves less in the role of trading advice for cooperation than in giving information that would help farmers avoid pollution and therefore prosecution. By and large, though, with responsive farmers the Pollution Inspectors did not adopt an overtly coercive stance, although elements of coercion were implicit in the nuances of the language they used, the resoluteness with which they pursued pollution problems, and the reputation of the NRA for prosecuting serious or persistent polluters.

With some farmers, however, Pollution Inspectors did adopt an explicitly coercive stance. These were the ones identified by the Pollution Inspectors as "problem" farmers because of their uncooperative attitudes. Such farmers were subjected to repeated visits, whether welcome or not, to underline the Pollution Inspectors' determination, and to keep the pollution problem under surveillance. The legal sanctions available were also explained, and Pollution Inspectors had few qualms about resorting to formal enforcement against obstinately unresponsive or obstructive farmers if the problem persisted. "There are always some farmers who don't care" commented one Pollution Inspector; another explained "You've got to have a legal comeback if you can't by friendly means get a farmer to do something".

Of course, individual farmers did not necessarily conform to the Pollution Inspectors' categories. A farmer initially categorized as persuadable or cooperative might eventually exhaust the patience of a Pollution Inspector through failure over a protracted period to take any remedial measures, leading to reclassification as a "problem" farmer. Informal approaches might then give way to the issuing of a warning letter. Prosecution in such cases was regarded as a failure of the negotiation process and of the Pollution Inspector's initial assessment of the farmer.

In the main, when major pollution was not involved and Pollution Inspectors therefore retained the discretion whether or not to initiate a prosecution, they tended to show considerable patience towards the farmers they regarded as persuadable. They were, for example, sympathetic to cases where farmers faced real obstacles to taking action but appeared to be doing what they could. Various considerations came into play. First, Pollution Inspectors did not want to jeopardize the relationship they had built up with the farmer in question. That relationship they saw as the basis for any real progress. By comparison, they regarded prosecution as an ineffective deterrent, seeing it mainly in terms of the financial costs to the farmer, with the level of fines generally being much lower than the costs of preventive measures. Pollution Inspectors acknowledged that the implicit sanction of prosecution could be instrumental in securing the cooperation of some farmers, but the actual deployment of the threat might poison relations.

Even when negotiations with a farmer had palpably not succeeded, and there seemed to be little to lose, Pollution Inspectors were concerned about upsetting the broader context in which they had to operate. The chances for a cooperative relationship with the individual farmer might look bleak, but any NRA response should not alienate local farming opinion.

One Pollution Inspector expressed strong misgivings lest ill judged prosecutions actually undermined the efforts to build up goodwill among local farmers, which he saw as the only realistic means to secure long-term improvements in pollution control. He contrasted cases that served to reinforce the NRA's authority – for example, cases of deliberate gross pollution, or cases where NRA advice had been flaunted – with ones that might undermine its authority. It would be imprudent, he thought, to take to court a farmer who had spent thousands of pounds on pollution control measures, as this could lead to bad publicity in the press, depicting the NRA as unreasonable, unless, that is, the pollution were very serious and roundly condemned by the same press.

Pollution Inspectors were also not indifferent to the compliance costs that farmers faced. Initial pleas by farmers that they could not afford any remedial action were usually discounted as standard reflex excuses. However, in cases where advice had been taken and the costs were likely to be particularly high and to cause real difficulty, there was a reluctance to proceed against the farmer. Here again, considerations of whether the action would be considered reasonable by the farming community came into play. Strong misgivings were expressed about strict adherence to the law that might lead to a farmer deciding to quit. One Pollution Inspector warned that it would be highly counterproductive if word ever circulated that the NRA had put a farmer out of business: "I don't want that"; but it could arise, he pointed out, if a small farmer were pressed to pay a lot for pollution control facilities. The subsequent rumours would always oversimplify the issue, but if farmers came to fear that the NRA would close them down, it could prompt "the wrong attitudes" and, worst of all, the "midnight valve phenomenon" of farmers deviously dumping waste into watercourses. This Pollution Inspector was "absolutely sure" that there had been a change in farmers' attitudes resulting in less pollution and did not want to jeopardize this by overly rigorous enforcement.

Another Pollution Inspector cited a case where strict enforcement would have involved closure of a dairy unit. The farm, identified as polluting on the farm campaign, was poorly located from the perspective of water protection. Situated in the floodplain between two rivers, it was "in effect in the wrong place". The only real solution to prevent pollution, which occurred as a result of the frequent flooding of the farmyard, was to close down the dairy unit. The owners had agreed to do this as long as they could move the dairy operations to higher ground and redevelop the existing farm buildings to offset their costs. However, they had been held up by delays in the granting of planning permission for the redevelopment, and the case had dragged on for over two years. Although the NRA could have initiated legal action, this Pollution Inspector felt that, as the owners were "doing what they can" in a very difficult physical location, there was "no point in prosecuting", despite there being a continuing problem.

In other circumstances, the gravity of the pollution caused meant that legal action was the first not the last resort. If the Pollution Inspector judged an incident to be major, a formal stance would be adopted. In the 1970s, incidents were dealt with in this way only if they were very severe and heavily publicized and, even so, the prosecution of farmers was specifically avoided (Hawkins 1984). The threshold that triggered legal action, regardless of the character of the polluter, had lowered significantly since then. By the early 1990s, indeed, farm pollution incidents were more likely to be prosecuted than were other types of pollution.[4]

The criteria used in the assessment of a pollution incident were a marked effect on the watercourse for several hundred metres, a fish kill, or a significant effect on someone else's use of water. In practice, Pollution Inspectors usually found it easy to diagnose a major case of farm pollution. The affected stream was likely to be green or brown in colour, and frothing, with a distinctive smell of animal or silage effluent. Dead or distressed fish might also be found. In the farmyard, tidemarks on ruptured or empty stores might indicate the scale of the spill.

Upon arrival at a serious incident in which the discharge was still flowing, a Pollution Inspector would take immediate action to stem the source. For example, if a lagoon was overflowing, the farmer might be told to dig some trenches to contain the waste or to fill a vacuum tanker and spread the waste onto suitable fields. As soon as possible the Pollution Control office would be contacted for assistance in investigating the extent of the incident, in instigating remedial measures, in warning water users and in collecting and photographing any dead fish. Remedial action could be a major operation. A small stream that had been polluted might be dammed to create a lagoon effect in order to intercept the head of the pollution. Spray irrigation equipment could then be brought in to irrigate the polluted water onto nearby fields. When a large watercourse was involved, aeration equipment might be used. Back-up of this sort allowed Pollution Inspectors, as the "investigating officers", to get on with their main task in such circumstances: the collection of evidence for a prosecution and cautioning the farmer.

Once a Pollution Inspector decided formal action should be taken with a view to prosecution, the procedure followed mirrored that used by the police: "anything you say may be used in evidence against you", the Inspector would warn the farmer. There would be no effort to be genial with the farmer, even where an informal relationship had been cultivated in the past. The Pollution Inspectors regarded legal proceedings as quite separate from and inimical to compliance-seeking strategies. They felt that once they had decided that legal action was appropriate "we have burned our bridges." The collecting of evidence in these circumstances was not seen as a means of reinforcing compliance but was carried out strictly to secure a prosecution.

The caution, indeed, marked the beginning of a procedure that Pollution Inspectors acknowledged had to be followed in meticulous detail if a pollution was to be successfully prosecuted. A key element was the taking of water samples to be used in court. To be admissible as evidence, these had to be in accordance with the provisions of the 1989 Water Act and were referred to as legal, formal or tripartite samples. The sampling kit, consisting of a bucket, bottles, funnels, labels and seals, always accompanied the Pollution Inspector. In describing the formal sampling process, Pollution Inspectors made repeated references to the need to avoid doing anything that would allow a defence lawyer successfully to challenge their evidence on a technicality, leading to the case being thrown out of court. They were keen to get the procedure right also because failure to do so might lead to disciplinary action from their immediate supervisors.

In collecting evidence, Pollution Inspectors were supposed to take a sample downstream of the discharge, then one of the discharge itself, then another upstream from the discharge so that the sampling procedure had no effect on the results. Equal care was needed in the splitting of formal samples, to provide one portion for the farmer, another for the NRA to analyze and a third to be stored away for future independent analysis if the NRA's analytical results were to be contested. The Pollution Inspector had to ensure the liquid was thoroughly mixed in the collection vessel before it was divided into three one-litre portions, preferably in the presence of the farmer. "Section 148 [of the 1989 Water Act] is crucial" stated one Pollution Inspector, emphasizing the point by quoting from it. He then explained the need to keep on filling the three bottles, even if they were overflowing. To leave some of the liquid in the collection vessel would divide it into four parts and this would contravene the letter of the law. Other Inspectors stressed the need to rinse the collection vessel carefully between samples, so that defence lawyers could not argue that one sample had contaminated another; and one detailed the deliberate and painstaking steps he took, to wash out collection vessels, which had obviously been of use to him in court. Each of the three bottles had then to be labelled and put into a bag and sealed with an official NRA seal. Once this had been completed, the discharger was allowed to choose one of the three bottles and was free to have the sample analyzed.

After the sampling had been done, the Pollution Inspector would take notes of the incident. The farmer would be questioned, although he or she had a legal right to remain silent. The Pollution Inspector also had a camera to collect photographic evidence. They would take photographs wherever they could, as this "helps magistrates to see" the impact of the pollution.

After a major pollution incident, a letter would be sent out advising the farmer that he or she might be prosecuted and asking for any comments on the incident. Pollution Inspectors prepared a case to be considered for prosecution from the analysis of the formal samples, the defendant's statement, details of the impact of the incident, any consent conditions and accounts of any previous problems the farmer had experienced. All this information was given to the area and regional supervisors, who made the final decision on whether the case should be prosecuted, in liaison with the region's Legal Department.

From this point on, the Pollution Inspector relinquished any control over the case. The NRA solicitor would take it through the court. If a defendant pleaded guilty, the solicitor would put forward all the evidence and present the case; Pollution Inspectors were needed only to give evidence if a case was contested. Most felt at a disadvantage in the legalistic atmosphere of the court. The validity of the evidence they had collected might be questioned and their sworn statements challenged. They felt that their expertise was not sufficiently respected by magistrates and defence lawyers; nor were they conceded the authority of their office as, say, a police officer would have been. As one Inspector noted, "in theory, I should be able to stand up in court and say I saw an illegal discharge", but in practice a Pollution Inspector's statement would not on its own have secured a conviction. Very few cases would be taken without formal samples, and those that were invariably had plenty of other evidence, such as quantities of dead fish. Legal action was thus often seen by Pollution Inspectors as an eclipse of their professional authority, for a number of reasons: because it signalled a failure of their preferred compliance strategy; because the case in question had been taken out of their hands; and because they felt their professionalism was not significantly recognized in the courts.

Conclusions

Charged to tackle farm pollution, NRA Pollution Inspectors had been armed with new legal powers and the moral authority of the environmental movement. However, most of their work with farmers was not the enforcement of pollution law but involved investigation of the occurrence and causes of farm pollution and efforts to persuade farmers to remedy the situation. Prosecution was treated as a last resort.

It was not that Pollution Inspectors did not regard pollution as wrong. On the contrary, more than any other group, they appreciated the threat posed by farm wastes to the vitality and utility of the water environment. But they found themselves charged with responsibilities, which meant that they had to work with the farming community. Without the cooperation of the large majority of farmers, their task would have been impossible. To bring to justice each and every farmer who had caused pollution of any significance would have meant prosecuting a considerable proportion of Devon's dairy farmers. But that would hardly have generated the commitment and the diligence – the required investment in waste facilities and the due attention to the management and disposal of waste – needed to address the problem. For, by their own inertia or indifference, farmers individually or collectively could exercise a sanction on regulatory authority imposed upon them. The potential sanction, moreover, was not only the negative one of inaction, but might also encompass, for the unscrupulous or bloody-minded farmer, the "midnight valve phenomenon". The impossibility of being omnipresent meant that Pollution Inspectors had to rely on the cooperation of farmers and this implied that they had to act in a way that the farming community regarded as reasonable. The law as a weapon of environmental morality thus proved to be a blunt instrument for regulators, who had to mobilize the sense of responsibility and of censure (i.e. the morality) of the farming community. Conversely, farmers needed to know how to manage farm effluents and, to be willing to take responsibility for this, Pollution Inspectors responded mainly by adopting another role, equally imbued with environmental morality – that of educators. We now turn to see how farmers understood the pollution issue and constructed their sense of responsibility for the problem.

Notes

1. Major, or "Category 1", pollution incidents are defined as involving one or more of the following: (a) potential or actual persistent effect on water quality or aquatic life; (b) closure of potable water, industrial or agricultural abstraction necessary; (c) extensive fish kill; (d) excessive breaches of consent conditions; (e) extensive remedial measures necessary; (f) major effect on amenity value (NRA 1994: 62).
2. Significant incidents are defined as those incidents that involve one or more of the following: (a) notification to abstractors necessary; (b) significant fish kill; (c) measurable effect on invertebrate life; (d) water unfit for stock; (e) bed of watercourse contaminated; (f) amenity value to the public, owners or users reduced by odour or appearance.
3. Most Pollution Inspectors were men.
4. For 1993, for example, agriculture was the source of 11 per cent of all pollution incidents, yet attracted 34 per cent of NRA prosecutions (NRA 1994).

CHAPTER SIX

The dairy farmers' accounts of farm pollution

Introduction

What did pollution regulation imply for the farmers? Pollution Inspectors were trying to persuade, cajole and, in a few cases, force them to manage their wastes in a way that would not cause pollution. To do so, the Inspectors had to engage with the farming community. But the farmers were not simply passive receivers of their exhortations and injunctions, and in any case had other preoccupations than water pollution. Indeed, just as the Inspectors sought to enrol the farmers to their view of pollution, so farmers in turn sought to enlist the Inspectors and others into an understanding of the difficulties and constraints they faced. During our detailed interviews with dairy farmers in Devon, many were keen to explain why the pollution issue was causing them such difficulties and to point to what they saw as the injustices of the recently introduced regulatory arrangements. This chapter draws on those interviews plus our participant observation of the interactions between farmers and Pollution Inspectors to provide the dairy farmers' accounts of farm pollution.

Dairy farmers are not, first and foremost, managers of animal wastes but producers of milk. In the period since the introduction of milk quotas, their strategies had altered from an expansionist and production-maximizing approach, to one of producing a fixed quantity of milk at least cost. However, stricter regulations coupled with high profile pollution prevention campaigns and the universal condemnation of pollution incidents meant that, increasingly, the "knack" of dairy farming was to produce near to the permitted quantity of milk, still at least cost, but at the same time ensuring that farm effluents did not get into watercourses. Waste management practices had to be fitted into an already busy schedule. We begin, therefore, with a description of the working routines of a dairy farm.

Life on a dairy farm

Although most farmers would be keen to stress that farming is as much a way of life as it is a business, the claim has particular pertinence for dairy farmers. Dairy farming requires that livestock be milked, usually twice a day, as well as bedded, fed and watered on a daily basis. It is this exacting routine with its close involvement with livestock that sets dairy farming apart in terms of the particular demands it places on family based labour.

For dairy farmers the day starts early. The cows are milked before breakfast, usually by the farmer but often with the help of other family members. The animals are familiar with the twice daily routine and wait, usually in the yard outside, and often in a recognizable pecking order, to enter the parlour. Once all the cows in milk have passed through, the floor and the milking equipment are thoroughly washed to comply with hygiene regulations. The washings from the parlour contain a mixture of cattle excreta, urine, spilled milk and the chemicals used to sterilize milking equipment.

Milking completed, the farm family and any hired hands share breakfast together in the farmhouse kitchen, typically between nine and half-past nine. The breakfast table provides the setting where the rest of the day's work can be discussed. The workload for the remainder of the morning will vary according to the time of the year, but among the routine tasks are the feeding, watering and bedding of the stock. In addition, there are other cattle to be looked after besides those to be milked, including young stock and replacement heifers.

The cubicles where the cattle are kept over the winter are cleaned daily, with the slurry either being washed or pushed out of each concrete-floored cubicle into a central channel. Waste management practices depend on the storage facilities on the farm (see Ch. 2). The slurry may either be pushed directly into a storage pit, pumped into a tank or pushed by a tractor with a blade up a ramp and into a "weeping wall" store. Some dairy farmers, most usually those with small herds, have no storage facilities for slurry and so push it up a ramp, and straight into a muck spreader.

Farmers whose animals are still kept in loose housing with straw bedding over the winter are able to clear the waste from their buildings far less frequently. Each day they simply add more straw and so the manure can be left, sometimes for up to several months, without being cleared out. A small minority of farms (just two in our sample of sixty, for example) have no livestock housing at all. The cows remain outside and have to be brought in from the fields twice daily all year round. Although such farms may be considered technologically backward, the problems they face in waste management are of a much lower order and are restricted to the washings from the farmyard and the parlour.

During the morning the milk lorry will call at the farm as part of its twice daily round to collect the milk. The farmer need not be present as the lorry driver will know how the system works. Other callers may include the vet, ADAS officials and sales representatives from agricultural merchants.

Once the stock has been dealt with the farmer may turn to fieldwork. If the weather is fair there may be the chance to spread some slurry, or perhaps to apply some nitrogen fertilizer, or even to spray some herbicide, to keep down nettles and other weeds.

Lunch is usually between one and two o'clock and this provides another opportunity to chat and catch up with the progress made on the day's tasks. Farm machinery or milking equipment may be in need of repair; buildings and field boundaries also require careful maintenance on live-stock farms; and, of course, there is the health of the animals themselves. A trip to the local market town may be required to fetch drugs from the vet, or machinery parts or building materials.

Waste facilities also require periodic maintenance. Settlement tanks and dirty water pumps have to be "de-clogged" of solid matter and gulleys and channels in the farmyard have to be kept free from obstruction. The *Code of good agricultural practice for the protection of water* urges that slurry stores be checked regularly and once a year be completely emptied and cleaned down to check for any signs of corrosion or damage (MAFF/WOAD 1991: 19). The Code also suggests that dirty water sprinklers should be moved regu-larly, surrounding land should be checked for signs of runoff or ponding, and tanks and lagoons should be examined frequently (ibid.: 32).

Any such tasks, along with any work in the fields, must be finished by mid-afternoon, because milking begins again between four and five o'clock. Depending on the time of year, there may then be a couple of hours of day-light left for some field work before the final rounds, when the farmer will ensure that all the farm animals are fed, watered and secure for the night.

Events in the farming year place added requirements on farm labour. Dairy farmers will often want to be present when cows are calving, particularly if there have been any problems during pregnancy. There is also the management of cattle breeding programmes, with cows increasingly being artificially inseminated, usually with the help of outside expertise.

In the summer months, when most cows are out in the fields, the daily chores around the farmyard and buildings may be reduced but the workload does not let up. The finer weather means farmers are able to get to grips with building improvements and the maintenance of the house, the yard, hedges and fences. Summertime also brings silage-making which is by far the busiest period in the dairy farmers' year (see Ch. 2). If the farm has a sheep flock, then lambing in the early spring is also a very hectic time; and for those farms with arable land, the harvesting of cereal crops in the late summer is an added demand.

The dairy farm can be viewed as a complex unit with family labour at the centre of a sophisticated technological system dealing with the daily pressures and constraints posed by the ceaseless round of milking, feeding and mucking out. Technological and management choices tend to be driven by the specific labour demands on the farm family in the context of this routine.

It will probably be necessary to draw on hired labour at some time during the season, such as for help with silage-making, or in the shape of a relief-milker to allow a holiday to be taken. (Many dairy farmers, though, do not regularly take a holiday). Hired labour may also be taken on, perhaps on a part-time, casual or contract basis, during particular stages of the family cycle, for example, if young children need to be looked after, or if a farmer approaching old age finds the farm work too physically exacting.

Equally, the shifting hopes and plans for the business – reflected in switches in farming enterprises and significant investment decisions – are inextricably linked to changing family circumstances and expectations. If the farm is being run with a succession to the next generation planned for, the farmer is more likely to build up the most profitable enterprises, will perhaps be more prepared to invest in new technology and buildings to make the farm easier to run, and will tend to manage and develop the business with a longer time-span than his or her own working life in mind. For example, plans to expand the dairy herd and reduce the sheep flock may result from the conjunction of changing trends in relative profitability and the need to create extra work to support a son or daughter entering the business. This can be contrasted with a middle-aged farmer who has no successor, who may be winding down the business by selling off or leasing milk quota and reverting to a less demanding beef or sheep enterprise, in order to reduce the workload in older age.

Life on a dairy farm

Of course, these daily and medium-term rhythms of the farm are subject to external factors too. As we saw in Chapter 2, dairy farmers operate in highly regulated markets and the changing political and economic relations of production pose significant constraints and pressures on their businesses. The regulated nature of the agricultural economy leaves farmers vulnerable to sudden policy changes. Since the introduction of milk quotas, farmers have come to feel that they live and work in a capricious world, subject to shifts in the political and economic environment as erratic as the weather. Many, indeed, bracket together the natural and political happenings beyond their control which disrupt their plans. Among the farmers we studied, it was repeated cuts in milk quotas and the spread of BSE that were proving the biggest shocks to be accommodated, although the establishment of the National Rivers Authority and the tightening of farm pollution regulations were also characterized by many farmers in similar terms. Indeed, a large national survey of dairy producers carried out in March 1991 found that pollution control regulation was the issue of most widespread concern (Centre for Agri-food Business Studies 1991). Seventy-five per cent of dairy farmers classed it as one of the "main problems" facing the industry (see Table 6.1).

Table 6.1 The Main Problems Facing British Milk Producers

Issue	Percentage of surveyed farmers considering the issue to be a main problem they face
Pollution control regulations	75
Break up of the Milk Marketing Board	67
Milk prices	55
Changes to the quota system	49
The dominance of a few large dairy companies	44
The GATT negotiations	43
Declining consumption of liquid milk	41
Greater competition from Europe after 1992	37
High interest rates	33
Increased milk sales through supermarkets	15
Declining consumption of butter	12
BSE	9
Others	8
High feed costs	5

Source: Centre for Agri-food Business Studies (1991).

By and large, the Devon farmers had come to accept that they should do something about farm pollution. Equally, though, they did not accept personal guilt for causing it. In part, this was because they did not think of themselves as necessarily to blame, but also because most did not consider farm pollution as a morally charged issue of right and wrong. In improving their waste management procedures and facilities they did not see themselves atoning for past misdeeds but as responding to external pressures on farming arising from changes in what society regarded as acceptable. They had a self-image of themselves as responsible people who respected the countryside.

Farmers were, therefore, offended by implications that they did not care for the countryside. They had an intimacy with it that others did not, and so criticism from "townies" was particularly unpalatable. After all, farmers' livelihoods depended upon an acute understanding of the natural and biological processes involved in the management of their land and stock. External interference in these matters challenged their self-esteem as well as their farming strategies. For many farmers, therefore, there was an underlying defensiveness and resentment in their response to pollution regulation.

Of course, the farmers saw the land and stock primarily as productive resources but most took a long-term view of their management of these assets, and most were not insensitive to the way their farming impinged on the countryside more generally. As one older farmer commented plaintively "farmers like the environment, otherwise they wouldn't be here". It was their efforts and those of their forebears in pursuing an honest living that had generated the pastoral countryside that modern society so valued. As another farmer put it, "it was farmers that *made* the countryside."

Farmers readily conceded that certain modern farming practices might be detrimental, but pointed out that they had been encouraged to adopt these by a government and a society wanting cheap and abundant food efficiently produced. There might, indeed, have been trade-offs between efficient food production and environmental conservation but the fact that they as farmers were being blamed for the consequences added further to their sense of being victims of an increasingly fickle public.

In the case of farm pollution, the public pressures on farmers were personified in the NRA's Pollution Inspectors. Inevitably, they were a focus for much of the farmers' resentment. Most farmers would have preferred to avoid their attention and some were decidedly brusque where encounters did occur. However, in a world in which it was felt that farming was widely misunderstood, Pollution Inspectors were not just another species of meddlesome bureaucrats but were influential officials who were obliged to enter the farmers' realm. Most farmers in their dealings with Pollution Inspectors sought to project themselves, just as they saw themselves, as responsible producers and countrymen. At the same time they took the opportunity to convey the practical and commercial pressures on contemporary dairy farming, and thereby to recruit regulatory officials to an understanding of the constraints faced in altering farming practices or investing in new equipment.

The environment and the ethos of production

Even though in an era of high-tech agriculture, the farmers would still express to us the need to conserve their farms' productive base. They liked their farms to be in "good condition", and most held that an important goal in their farming lives was to "improve the farm" so that it could be passed on "in a better condition". By this they meant that the accounts should be sound and that the equipment in the milking parlour, the tractors and field machinery, the farm buildings and perhaps the pollution control equipment should all be in good working order. But besides the technology and the layout of the farmyard, which helped to make the farm *easier* to manage (a crucial consideration given the dependence of most dairy farms on family labour), conservation of the farm's land resources remained a crucial element in the farmers' understandings of their farm's "condition".

They expressed some of these notions in terms of their soil, or land being "in good heart". This concern had two strands. First, field boundaries had to be properly managed and kept in good repair, thus ensuring not only that hedges, ditches and fences were secure to the stock but also that the farm looked tidy and well kept. The second strand involved the maintenance of soil fertility, and arose primarily from dairy farmers' concern for their productive asset, not only for their own use but for their descendants' also. This was often expressed in terms of responsibility to the land, as in remarks to the effect that a good farmer "should not take more out of the soil than gets put back in".

This utilitarian notion of resource conservation, set within a logic of farm improvement, was reinforced within the farming community. The farmers were asked if they made comparisons between themselves and their neighbours, and if so, on what basis. For the four-fifths who did so, the bases of comparison fell into three main categories. The first group (comprising 30 per cent of these farmers) referred to notions of good farming, the health of stock and tidiness of fields. One farmer explained [I look at] "not how much money they make, but how they're keeping the farm – what sort of husbandry". A second group (30 per cent) made comparisons mainly based on the timing and methods employed on the farm, such as who was cutting silage first and what types of machinery were used. The third and largest group (40 per cent) made comparisons in terms of yields. A typical response in this group came from a farmer who said, "I always try to do that little bit better, to get a better yield". Overall, therefore, production-maximizing values continued to have strong currency among the farmers, although notions of good husbandry were also important.

How, if at all, did these resource conservation preoccupations of farmers articulate with the environmental concerns of the non-farming public? By the early 1990s, farmers generally had become sensitized to environmental issues. Eighty-three per cent of those interviewed acknowledged that agricultural practices could have an adverse effect on the environment. In all of our discussions with them, however, only two expressed misgivings that farm pollution might damage the farm's productive base. For the rest, environmental concerns lay outside the logic of farm improvement.

For both dissenters, their misgivings centred on low rate irrigators used to spread dirty water on land which they feared was killing worms essential to the health of the soil. One of the two was a young farmer running an 80 cow dairy herd on a 100 ha mixed farm with a loose housing and farmyard manure system. He expressed a distinctly "non-productivist" attitude to farming: "I'd like to think that I farm 80 cows *well* rather than increasing for the sake of it. I get more pleasure from doing it right rather than getting bigger. Farming has to be a pleasure to me first. I can't do anything for profit alone". He had a clear sense of what passing on his farm "in a better condition" would mean, namely "without any damage and still in a healthy and farmable state, leaving nothing that the next generation would have to work hard at to put right". In this regard, he saw the current fashion for dirty water systems as not only damaging to the land but also as an ineffectual solution to farm pollution. "They just pump the effluent around the farm, onto the land and it gets back into the drains". He had considered moving over to a cubicles and slurry-based system but the potential slurry problems had dissuaded him.

To this farmer, it was quite conceivable that modern agricultural practices were damaging farming's productive base. The view of the vast majority, however, was expressed by one who remonstrated that "No farmer is going to do something that is going to ruin his own property and his own livelihood".

Farmers and farm pollution

The vast majority of farmers thus did not accept farm pollution as an indictment of their farming systems. This did not mean that they denied the need to improve matters, but in taking action they saw themselves essentially having to respond to external pressures, arising from increased public sensitivities.

Two thirds of them acknowledged that environmental concerns had begun to influence the way that they farmed. Responses ranged from using environmental criteria to decide on a whole production system at one extreme, to "feeling a bit guilty when I use nitrogen and sprays" at the other; with over one third of the sample (unprompted) citing the more careful handling of effluents to minimize pollution risks. One of the larger farmers explained how environmental considerations had been an important factor in reorganizing the cropping system: "We now put down autumn cover crops and have stopped applying nitrogen in the autumn. Also, we are much more aware of how to apply slurry. We have done some trials with injecting slurry". Practices once widespread, such as letting slurry run straight into streams or ditches, were now frowned upon.

Some farmers clearly felt coerced to act. A farmer in his early 40s with a medium size holding remarked: "You hear of people in the village being prosecuted, and more people than ever are putting in dirty water systems now. You hear that the NRA are hot after everyone who's polluting". One heavily indebted farmer had recently spent £6000 that he could ill afford installing pits and a pump to handle dirty water from his farmyard: "The NRA visited us several times last year", he explained, adding "they were rather persistent". Others, however, saw the need "to put agriculture's house in order"; for them it was a question of farmers' social responsibility, of farming's public image. For example, when asked what he thought about a strict prosecution policy for farm pollution incidents,

Figure 6.1 A mucky farmyard. (Photograph courtesy of the
Water Services Association)

one farmer said "there's nothing wrong with it. The pollution has got to be stopped". Whether out of a sense of being coerced or one of social responsibility, however, farmers felt they were responding to outside concerns. There was a paradox here in the very meaning of farm pollution to the farmers which reinforced the sense among them of being subject to external and obtuse pressures. They spent their working days surrounded by animal wastes but difficulties only arose when someone else was affected and complained.

Indeed, farm wastes were a ubiquitous fact of life for them. More often than not, the farmyard, which usually adjoined the farmhouse, was splattered or caked with manure. Each morning as they began the working day, farmers stepped into a world of muck (Figs. 6.1 and 6.2). It caked their boots, dirtied their clothes and stained their hands. Perpetually, slurry had to be cleaned from buildings, pumped into tanks, spread on land, scraped from boots and washed from clothes. Undeniably, it was farmers, as well as their workers and their families, that were most exposed to farm wastes, their unpleasantness and risks, to the point of extreme familiarity. The dangers were not to be underestimated: noxious fumes could collect in effluent handling equipment, and there had been cases of farmers being killed while trying to break up the solid crust that often formed on top of stored slurry. But no-one counted any of this as pollution, certainly not the farmers. For the farmer the problem was not so much one of causing pollution as being detected doing so. Although they constantly experienced farm *wastes*, for many of them their initial experience of farm *pollution* was a visit from the NRA or a complaint from a neighbour.

The farmers' very familiarity with farm wastes inclined some of them to question the significance of farm pollution as a threat to the environment. Four-fifths of the surveyed farmers felt that it was a less serious threat than pollution from industry. A few of the farmers elaborated on this point drawing on notions of farm waste as natural substances. One said "I feel that chemical waste [from industry] is much more serious compared to farm waste. I know it kills fish, but I can't see what's wrong with brown water – it's natural. I know silage effluent is bad, but it's only like juice from grass. It's not like liquid from industry". Another farmer commented "Industrial pollution can be a lot worse – they're using chemicals. Our pollution is a natural pollution. Animals have been making it for years. Industrial pollution is modern pollution. It's not like country smells and that". While these farmers emphasized the biological nature of farm

Figure 6.2 Slurry being stored in the corner of a farmyard before being spread on land. (Photograph courtesy of Louise Morriss)

waste, others asserted the naturalness of farm pollution through its meteorological origins. A number, for example, pointed out that, as pollution was most likely during heavy rain, any runoff or spillages would be rapidly diluted and washed away by rivers in spate. The implication was that any ill effects were essentially transient. One young farmer (in his early 30s) running 70 milking cows on 60 ha explained: "I try to stop pollution in every way. I think most people would. But it's impossible when there's really heavy rain and it washes off the yard. I don't think it does too much damage because it's during flooding conditions and it gets diluted and washed away."

For most farmers, the disjuncture between their own quotidian experience of farm waste and pollution regulation as an unexpected visitation firmly placed the public phenomenon of pollution into the category of external capricious event. In their own eyes, the fact that the immediate cause of the problem might have been a sudden downpour, prolonged rain or equipment failure cast them as hapless victims rather than perpetrators. Repeatedly farmers referred to such incidents as "acts of God". Much farm pollution was portrayed in this way as accidental; therefore the farmers involved were not necessarily culpable. In a telling turn of phrase, one young tenant farmer with 60 dairy cows protested: "Accidents can happen. If a slurry store breaks and the farmer can prove it was an accident, then it's wrong that his name gets dragged through the mud". This inversion of the idiom of environmental morality neatly captured the indignation of many farmers lest pollution besmirch their good character.

While the immediate causes (and consequences) of farm pollution might be presented as natural ones, the public and official responses were understood to arise from changes in society. This was how farmers made sense of the fact that they had always had to dispose of farm waste but it was only in recent years that they had been challenged about the way they did it and the consequences. Basically, public attitudes were seen to have changed, and the climate of opinion was now perceived to be much less sympathetic towards farming and less tolerant of the problems farmers faced.

Of course, farmers recognized that modern farming methods such as slurry systems and silage-making involved a higher risk of pollution, but most of them in the survey had adopted these methods several years earlier. They had done so, moreover, with official blessing. It was therefore unfair, they felt, for them to be blamed. "It all boils down to the pressure to farm intensively", one farmer reasoned. Another in his early 60s, with a 50 cow milking herd on 32 ha explained: "It was the move away from mixed farming. We have been taught to specialize and that means slurry problems. Mixed farms didn't have waste problems". A third of the farmers explicitly blamed "the system" in this way.

If farm pollution was an unfortunate consequence of farming methods that the Ministry of Agriculture itself had promoted, then the wider community with its demands for plentiful, hygienic and inexpensive milk, must shoulder some of the blame. Did not the 50 per cent grant from public funds towards pollution facilities acknowledge that complicity? For many of the farmers interviewed, this was the real source of injustice – that in finding the finance for the other 50 per cent themselves they were being expected to pay out considerable sums to comply with a change of policy. As one farmer said indignantly, "If they're going to enforce something like that, they should cough up the money".

Farmers did not deny all personal responsibility for farm pollution. When asked who specifically was to blame, two-thirds of them spoke of a small number of "bad farmers" or "cowboys" who deliberately emptied their wastes into rivers. This group was the counterpoint to the majority. Their wantonness could be contrasted with the responsibility of the majority. Their culpability underlined the misfortune of those caught "accidentally polluting". Such comparisons were frequently made when farmers were asked their opinion on prosecuting cases of farm pollution. With regard to "the cowboys" there was general condemnation and an attitude of "they get what they deserve". They should be prosecuted not least because they tarnished agriculture's good name. One farmer in his late 50s running a very small farm of 17 dairy cows on 17 ha of land explained how a stricter prosecution policy was the best means of dealing with these "bad farmers": "It would stop a lot of them. It would stop the 'don't care' attitude of those who let stuff get into the water".

There was, however, wariness or opposition towards prosecution in other cases and the distinctions drawn revealed that for most farmers culpability was wholly to do with intentionality, with deliberate acts of wickedness. The tenant farmer quoted earlier protesting the injustice if a farmer's name got "dragged through the mud" for accidentally causing pollution added equally forcefully: "If stuff gets tipped in the river on purpose then they should lock him up because that's not on". Another, older farmer running a 90 cow milking herd stressed that "the law's the law, but there's always that odd occasion when there is an accident like an unforeseeable burst. Even with slurry stores, they have a life span. You get these natural accidents. They happen". One farmer drew the distinction this way: "It's difficult to regulate strictly against accidents, when they're not intentional. If guidelines are clear and strict and then are blatantly broken, then a strict prosecution policy would be OK. But it's a bit like the difference between consciously speeding in a car or just driving along unaware with a dodgy brakelight". Concern was also expressed that, even though farmers might have done all that was expected of them in improving their facilities, "accidents" could still happen. Investment in equipment could never provide "a hundred per cent guarantee against pollution".

A third of the farmers argued more generally that prosecution was the wrong approach to an already beleaguered sector. One such farmer, in his 60s and running a fairly large farm with about 120 dairy cows, complained, "It will drive some farmers out of business. There won't be no more farmers . . . Agriculture is being butchered in this country. We've increased efficiency and have done what we have been told to do, and now it's almost as if people don't want a farming industry". Another farmer, a local magistrate, explained what he saw as the limit of a strict prosecution policy: "I know people who have been prosecuted . . . I don't think it's the best way. You're taking money off the farm for the fine. That money would be better spent on putting in the [pollution control] equipment. It's better to keep it out of the courts, to talk to people rather than prosecute."

In summary, although there was greater sensitivity to environmental issues among the farming community, environmental concerns were still generally seen by the farmers very much as external pressures on the farming sector. In the main, the changes in farming practices that were occurring were not arising as a result of farmers acknowledging some past "errors" of their ways, nor from fears that environmentally harmful agricultural practices might actually damage the productive base of their own farms. Rather, farmers were acquiescing in environmental pressures, either because they felt coerced to do so, or from a sense of social responsibility. Linked to this were farmers' perceptions of where the responsibility lay for farm pollution. Deliberate discharges of effluents into water should be strictly dealt with, but for other types of pollution incident, farmers stressed either their accidental nature or the fact that farming practices and technologies had been strongly shaped by government policies.

Pollution control and the logic of farm improvement

In most of their dealings with farmers, Pollution Inspectors were trying to get them to alter or improve their waste management. But for the farmers, as we described above, waste management was very much a secondary chore to the serious business of efficiently producing the right quantity of wholesome milk. This primary purpose demanded close attention to animal husbandry, including the health of the cows, the management of their lactation cycle and their feeding to ensure sustained milk yields. One aspect of waste management was seen as crucial to milk production and that was dairy hygiene. It was vital that the farm's milk should be fit and wholesome and so the parlour had to be kept scrupulously clean. Effective application of slurry could also reduce the need to buy in fertilizer but the slurry was very variable in its fertilization values. Otherwise, sound waste management contributed little, if anything, to the financial efficiency of the business, but was a charge upon it in terms of time and money. It would therefore not normally have been a major priority for farmers in planning improvements to their farms.

For most farmers, indeed, straight investment in pollution control tended to be seen as "dead money". A slurry store or a dirty water system contributed nothing to the commercial viability of the holding. As one dairy farmer complained in the *Farmers Weekly* (12 April 1991, page 70), "The money spent on pollution control does not generate one penny of extra income and the rules are being tightened at a time when incomes from dairying are falling and there is great uncertainty about quota levels". Moreover, farmers were unable to pass on the costs of improved pollution control to their customers through higher prices. As a Country Landowners' Association spokesman explained, farmers "can't write to the Milk Marketing Board and ask for so much more per litre of milk to cover the cost of anti-pollution measures taken" (*Farmers Weekly*, 12 April 1991, page 70).

In these circumstances, the 50 per cent grant then available at the time of our survey for pollution control facilities was considered by farmers to be both necessary and justified. However, many farmers felt that the level of grant should be higher. One older farmer explained, "Where do you get the money from when there is no return on the investment. Even if you get a 50 per cent grant, you've still got to find the money first". Another farmer complained that, three months earlier, he had spent £27 000 on new pollution control facilities which had been approved for grant aid but the cheque had not yet arrived and in the meantime he was having to pay the interest on the loan himself at a time of high interest rates.

These financial complexities and costs meant that pollution control investments needed to be fitted into the farmer's investment horizons. Factors such as the availability of spare capital and the changing long and medium-term plans for the farm would often come into play in deciding what steps to take to install new equipment and address pollution risks. For those farmers approaching retirement age and with no plans for passing the business on to the next generation, there was often a marked reluctance to spend thousands of pounds on equipment which would yield little in terms of the improved profitability of the business.

For others, however, investing in pollution control could be justified within the terms of their own logic of farm improvement and often in these cases the eligibility for 50 per cent grant aid was seen in a different light, as a useful incentive. Improved effluent storage and pollution control facilities might help make the farm easier to run, enhance its capital value or facilitate future expansion. As an article in *Farmers Weekly* (17 January 1992, page 51) stressed, on many farms "the installation of a grant-aided new or improved slurry store together with a dirty water disposal scheme will be a once-in-a-lifetime opportunity to improve working conditions around the dairy unit, reduce the labour required for slurry handling, and at the same time increase the value of the farm" (Fig. 6.3).

Several of the farmers we interviewed explained how new slurry stores or dirty water systems were making their lives easier. For example, a new store would mean that slurry would not have to be routinely spread on land every day, thus freeing up time for other tasks about the farm. Storing slurry and choosing when to spread also made for better use of its nutrient value. Finally, new facilities might mean that it was easier for a farmer to handle additional stock, perhaps after leasing or buying extra milk quota. Several farmers had taken the opportunity provided by the grant not only to improve their waste handling and storage facilities but also to enlarge them, to allow for a possible expansion of the dairy enterprise at a later stage. Strictly speaking, this broke EC rules against the use of grant aid to expand farm productive capacity.

Some farmers would have liked the grant to cover a greater range of items. Expenditure on the roofing of yards, for example, was not eligible. In many cases, roofing would have provided an effective means of improving pollution control because rainwater could be kept separate from farm

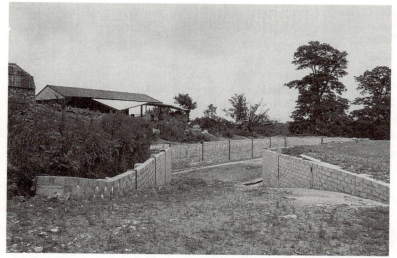

Figure 6.3 A large concrete-walled slurry pit under construction. (Photograph courtesy of Graham Cox)

effluents greatly reducing their volume and so increasing the length of time they could be stored before spreading. Several farmers expressed frustration that they could not obtain grant aid for what they saw as a sensible solution to the pollution problems on their particular farms.

Most of the changes in farm effluent management remained within a dominant food production logic. Some farmers, indeed, admitted that "more careful" use of slurry or nitrogen was because financial pressures were forcing them to be more efficient with inputs, rather than from any feelings of conscience towards the environment. More generally, quotas and subsidy cuts aimed at curbing over-production were seen to favour the boosting of margins through increased efficiency of input use rather than the pursuit of maximum yields.

Farmers and the NRA

Most farmers felt that they were doing their best to manage farm wastes and prevent effluents from contaminating watercourses, often in difficult circumstances. Nevertheless, even on a well run farm, pollution might happen and bring with it the unwanted attention of the NRA. To many

Figure 6.4 An NRA Pollution Inspector discussing a new waste storage system with a farmer. (Photograph courtesy of the Environment Agency)

farmers, the NRA's Inspectors were pollution personified. In particular, they seemed to embody that shift whereby society had come to value a pristine water environment over and above its vital food supplies. As one farmer remarked of the NRA: "They're tossers. They're not practical people. I don't think they understand. They go in with their water samples and they want everything to be perfect. They don't give a damn about what goes on on farms or what the implications are for farmers."

Of course, no farmer would want to be caught polluting. But in a capricious world in which farming was subject to distant and anonymous decisions, Pollution Inspectors were not regarded simply as obtrusive and overwheening officials. However unwelcome their message, they were not faceless officialdom. They entered the farm and attempted to strike up a relationship with the farmer (Fig. 6.4). Their very presence, which was so disquieting or threatening to some farmers, provided an opportunity to others to draw the Pollution Inspectors into their world, to explain the constraints under which farmers had to work and to implicate the Pollution Inspectors in the decisions and actions that needed to be taken. In short, the Pollution Inspectors were a crucial connection to that wider society in which farming was so misunderstood.

Undoubtedly, though, there were a number of farmers who hoped that by "keeping your head down" they would avoid the attention of the NRA. Of the 60 dairy farmers surveyed in Devon in 1991, 25 had not upgraded their pollution control facilities since 1981. Six of these had plans to improve pollution control and were in the process of seeking advice or applying for grants, leaving 19 farmers who had made no changes and had no plans to do so. While a few of them were running lower risk, non-slurry systems, some of the others admitted they were "lying low".

One such farming couple, who were in their 40s with two teenage children, had taken over the tenancy of a 55 ha farm from his father in the early 1980s. They had subsequently expanded the dairy herd from 40 to 50 cows but the waste facilities remained primitive. The cows were housed in cubicles but there was no storage for the slurry which had to be spread daily. Likewise, none of the dirty water was collected, despite the fact that the farm was close to a river. The farmer's wife mused, "I suppose we must be polluting, and our dirty water probably does go into the stream, but it's a tributary. It's not as if it's the main river itself". Most of the neighbouring farms in the valley had been visited by the NRA, but not yet theirs. The couple talked with evident dread of the possibility of an impending visit, commenting "Our days are numbered". They were very worried about what types of spending might be called for in harsh economic times – the farm's viability had tailed off sharply in the previous two years. They joked that their strategy would be to "hide behind the sofa" when the NRA finally called, although the wife did concede that it would be a relief to find out what changes might be required and to put their minds at rest. It had not occurred to them to approach the NRA themselves.

The vast majority of farms had received visits from NRA officials. These visits fell into three categories. First were the NRA-initiated "farm campaign" visits to all farms in a locality to check on waste management facilities and practices and to discuss any potential problems and possible improvements to reduce the risk of pollution. Second were the farmer-initiated visits where the NRA would be called onto the farm to discuss plans for the installation of new pollution control facilities or to secure NRA approval as part of the farmer's application for Farm and Conservation Grant Scheme funding. Finally, there were those visits by the NRA in response to a pollution incident. The pollution might have been reported by a member of the public, or by the farmer, or might have been spotted by an NRA Inspector.

Farmers' accounts of their encounters with the NRA varied according to the circumstances. We have supplemented their accounts here with evidence drawn from our participant observation of Pollution Inspectors' farm visits. A complaint of several farmers was that NRA officials had been brusque and unhelpful in investigating pollution incidents. Accounts of what had happened were often prefaced with details of how a pump or valve or a containing wall had spectacularly failed and how the farmer was struggling to stem the problem when the NRA official arrived on the scene.

One farmer's dirty water pump had failed in heavy rain leading effluent to spill over into the stream that ran alongside the farmyard. The farmer was trying to pump out the dirty water from the area near the stream when the "despicable little man" from the NRA arrived: "The first thing he said when he arrived was 'anything you say may be taken down and used in evidence'. He took up half the day taking samples from the river. This approach gets farmers' backs up".

Farmers contrasted the disposition of NRA staff investigating an incident with the approach adopted on other occasions. For a number of the farmers, the pollution incident had involved a different NRA official from the one who had paid them a farm visit. These farmers tended to personalize the distinct stances adopted, rather than dwell on the different circumstances. One young farmer running an intensive dairy farm with inadequate waste management facilities commented as follows on his involvement with the NRA: "There's one bloke, you couldn't meet a nicer bloke. He advises you and is helpful in any way he can be . . . He's somebody you could take to straight away . . . But the other bloke is a right one." When the farmer was reported for causing pollution by his neighbour, it was the "other bloke" who came to take the water samples: "He was ignorant beyond words, and very unhelpful". The farmer concluded: "They've all got a job to do, I suppose, but if they could just be nice about it. It makes a lot of difference".

The farmers preferred a cooperative and understanding relationship with NRA officials and were put out by what they regarded as officiousness. They did not find the persona of the law enforcer acceptable. Of course, farmers' feelings about being investigated over a pollution incident were coloured by the threat of prosecution hanging over them, but they also found it demeaning to be treated brusquely. At such times, NRA staff were seen as unyielding and punitive officialdom.

Under other circumstances, of course, an NRA visit might be much less distressing, but not necessarily so. All dairy farmers, including those who had never been at the centre of a pollution incident, were acutely aware that NRA staff "carry a big stick". One farmer commented: "the local NRA bloke, I know him. He's abrasive. He thinks he's God and he struts around as if he owns the place. He could be a bit more civil. He's very policeman-like. Although most policemen I know are better than him".

The message from NRA staff when they came to inspect a farm's facilities might also be quite unpalatable – for example, the need to spend thousands on pollution control. Even so, the manner of the approach was different and not so seemingly devoid of sympathy as when a pollution incident was being investigated. It was not just a question of courtesy but of the scope for the farmer to engage officialdom and thereby to avoid being merely the object of unbending regulation.

Some farmers, indeed, expressed considerable alienation at just this: at being at the receiving end of apparently capricious and insensitive rules. One farmer, for example, expressed contempt of both the NRA and the agricultural advisory service, ADAS: "The NRA pass you to ADAS. ADAS pass you to the NRA. I don't think any of them know what they're on about." He went on, "you can't get any sense out of them. The pollution regulations keep changing and the goal-posts keep moving . . . You could do just what they say and then the next minute you're caught out because they change the regulations. Anyway, even farmers who have done all they say have still been prosecuted".

Most farmers, therefore, took the opportunity of their involvement with NRA officials to seek to draw those officials into the decisions that had to be taken. This included conveying to them the constraints under which they as farmers had to operate: the risk to flooding of a farmyard, the lack of capital, the volume of waste, the difficult circumstances of dairy farming. It also involved trying to domesticate regulatory authority. As we saw in Chapter 5, Pollution Inspectors were adamant that they were not there to advise farmers, but this was precisely the role into which they were cast by the farmers. Farmers did not simply want to be informed that they risked causing pollution but what steps they should take to avoid doing so. For many this boiled down to divining what needed to be done to placate the NRA. When asked what their most important source of advice regarding waste management was, most farmers cited the NRA. Our participant observation revealed Pollution Inspectors giving a range of types of advice, going well beyond the expected advice related to possible risks and sources of pollution on the farm, and the regulations covering the storage of farm wastes. They also gave advice, for example, on the operation and maintenance of waste equipment and on spreading practices and procedures (see Ch. 5).

Many farmers had found that the Pollution Inspectors were unwilling to be drawn on certain types of advice, particularly the technical details of what pollution control facilities might be needed. This apparent obtuseness was a source of considerable frustration for the farmers, not least because many of them saw installing the correct technology as the guarantee that their pollution problems would be solved. Such an attitude was reinforced by the Farm Waste Regulations and the Farm and Conservation Grant Scheme. The final reassurance sought by farmers was the blessing of the NRA, but all too often NRA staff would not commit themselves on what needed to be done. One such farmer was planning to expand his herd from 90 to 140 milking cows and develop a new dairy unit further away from the village. As part of the development, he had recently installed a whole new pollution control system, including a "weeping wall" store, new channelling and a dirty water system of settlement tanks and a low rate irrigator at a total cost of £45 000. Before proceeding, he had called the NRA onto the farm to discuss his plans and had asked whether they thought them suitable, but their response had been "let's wait and see when it's put up". The advice, he complained, had been "wishy washy", but "when you are investing that sort of money you want to know whether it's right. They haven't come up with the answers one way or the other". A similar complaint was expressed by a small dairy farmer (with fewer than 40 milking cows) who had become worried about more stringent water pollution controls and so had invited the NRA to inspect his farm to see if any changes needed to be made: "they were extremely vague and wouldn't commit themselves at all. They said 'We are not an advisory service. See ADAS' . . . We'd like to know exactly what we should do, but the NRA wouldn't say. We were trying to be responsible but we came up against a brick wall." This sort of stance was regarded by farmers as a sign of aloofness or indifference towards the practicalities of farming.

Farmers, however, were not inclined to leave the matter there. They expected experts to provide certainty and reassurance. The Farm Waste Regulations required the NRA's assent to any material change in a farm's waste storage facilities, and such assent was needed if the farmer was to apply for grant aid. Universally, farmers referred to this assent as the NRA's "approval" or "seal of approval" and even the Pollution Inspectors acquiesced in this terminology. But their complicity often went much further than that. It was not unusual for Pollution Inspectors to help farmers to fill in the forms through which NRA assent was sought. In the case of Farmer Shields, reported in Chapter 5, the Pollution Inspector actually filled in the form and drew the required sketch map for the farmer for a new dirty water system that he proposed to install.

Once the work had been completed, farmers tended to be even more keen to gain the blessing of the Pollution Inspectors to the improvements made. Farmers then were usually at their most welcoming, and often at their most open. It would seem that they saw the Inspector's approval at this stage as a sort of guarantee, if not against pollution occurring, then against NRA retribution if it did. This tactic was particularly transparent in a number of cases we observed where the farmers were keen not only to display the completed works but also to draw the Inspector's attention to any apparent defects. Evidently, these farmers were seeking not only approval for what had been done, but also complicity in any of its deficiencies.

One farmer, showing a Pollution Inspector around his recently completed, grant-aided, slurry management scheme, repeatedly pointed out what he considered to be potential defects of the concrete base. The contractors, he explained, had needed to dig out several feet of boggy matter and put in a large amount of hard core before the new concrete yard and slurry store could be constructed. He was concerned about settlement of the concrete and recalled that the first slurry put into the store had disappeared overnight. He drew the attention of the Pollution Inspector to joints in the concrete that were not well sealed and to signs of seepage. He then speculated on whether any seepage would lose its polluting power if it were to percolate through the soil and reach a brook which was 60–70 metres away. The farmer was clearly seeking reassurance. The Pollution Inspector suggested the use of a sealant but, as to whether the brook would be affected, he remarked "we will have to wait and see". Tellingly, the farmer immediately responded "if there's any problem, I'll be on to you". (We analyzed this interchange from the Inspector's point of view in Chapter 5.)

Farmers felt that NRA staff, through their involvement in regulating waste management facilities, were implicated in the actions taken. This was most clearly expressed by a farm family who had recently been at the centre of a pollution incident. They had been "plagued" with pollution problems on their 100 ha farm with over 100 milking cows. A four-tank settlement system with a dirty water pump had recently been installed, but there had still been pollution problems. A channel in the farmyard took the slurry from the yard to the collecting pit. Just a week before our interview with this farmer, after several dry days, the channel had become blocked with about a week's worth of dry scrapings. Following some heavy rain, the rainwater from the adjacent road and the farmyard built up behind the blockage. Once it broke through, there was a rush and the dirty water system was overwhelmed. The pump stopped working and the effluent spilled over into the stream that ran alongside the yard. The farmer had done his best to contain the spill but the NRA had been alerted and had formally investigated the incident. The farmer complained, "We put in all that's needed. First there was the slurry. I invited the water authority to come up and discuss it and they said 'put in a big store' so we did. Five years later, that wasn't enough. They wanted a dirty water system, and so we did that. We've done everything they asked us to do . . . I feel tremendously let down, especially if the NRA [decide to] take legal action".

Conclusions

The Devon farmers we met had become sensitized to environmental concerns and recognized that some farming practices could harm the environment. But they did not see farm pollution as necessarily an indictment of the way they farmed, although they were keen to avoid causing it, either out of a sense of social responsibility or from wanting to stay on the right side of the law.

Their self-image was that of responsible producers and countrymen, but they were at pains to point out that they were also subject to forces and constraints beyond their control. On the one hand, through public demands for plentiful supplies of milk efficiently produced, they had been pressed to adopt farming practices that posed greater risks to the environment. This implied, at the very least, a shared responsibility with the wider society for the consequences. On the other hand, as farmers they were subject to the uncertainties of the elements and many of the cases of pollution that resulted were portrayed as unfortunate accidents.

To blame society or nature in this way might seem commonplace responses to charges of wrongdoing or error, and in that sense the farmers were not unusual. What was particular was the way in which they could draw upon a natural morality – to justify themselves and explain the constraints upon them – based on the elemental nature of agriculture. Thus, on the one hand, in supplying its primary sustenance, society had a basic and absolving dependence upon agriculture, but, on the other hand, the farming of land and animals had an unavoidable dependence on nature. These verities gave the farmers a strong sense of where they stood in the order of things. However, whilst pollution could thus be attributed to the capriciousness of nature and society, it also implied a disturbance to that order which was deeply unsettling, if not threatening, to their identity and sense of worth. Their reaction was to reiterate their own feelings of personal responsibility and attachment towards the land and the countryside and to denounce those who wantonly damaged the environment and thereby tarnished agriculture's good name.

In the next chapter, we go on to examine the process of negotiating pollution control on farms in more detail, looking at the way that different groups of actors represented the nature of the problem, and the networks of inter-relations within which they were embedded. The dairy farmers' accounts have highlighted an increasingly key element in the story. It was not just the NRA that "imposed" pollution regulations upon farmers. Other people in the countryside were also willing to challenge what they considered to be unacceptable practices. Therefore, in Chapter 7 we go on to explore farm pollution in the context of the changing rural order.

CHAPTER SEVEN

Pollution control and social networks

Introduction

It was in the interacting working lives of farmers and Pollution Inspectors that farm pollution in Devon was caused, detected and sometimes rectified. The pollution issue was constituted by the reciprocal efforts of the Inspectors and the farmers to recruit the other to their own outlook on besetting problems. In trying to fix one another, however, they were separately involved in additional networks: the farmers were part of an agricultural community living and working in a changing rural world; and the Pollution Inspectors were part of a large, new institution with a prominent public profile and extensive dealings beyond the farmyard. These other networks were potentially a resource to be drawn on, but might also act as constraints; through them, other actors could be called upon, or in turn, might seek to impose their own perspectives. Two such prominent sets of actors influencing the constitution of the farm pollution problem were agricultural advisers and environmental activists. In this chapter our interactionist analysis is extended to embrace these other actors, drawing upon local research "shadowing" ADAS officials in their work advising farmers on how to deal with farm effluents, and upon interviews with local environmental activists campaigning against water pollution. Below we examine, in turn, their respective roles in the farm pollution issue, before returning to refocus on the farmers and the Inspectors and on the different networks in which they were both embedded and in which the fate of the water environment was sealed (see also Seymour et al. 1997).

ADAS, farmers and farm pollution: the technical discourse

Farm pollution control was undoubtedly seen as a threat by farmers, but not only by them. ADAS, the Ministry of Agriculture's official farm advisory service, had, in the years before milk quotas, advised farmers to switch to slurry systems, to enlarge their animal housing, to concrete over yards, to under-drain fields and to make silage rather than hay. Such single-minded expansionist advice had taken a knock with the imposition of milk quotas, but rising problems of farm pollution further discredited it. The damage done to the legitimacy of ADAS paved the way for the service to be cut and commercialized in the late 1980s. Against this trend, the shift in policy towards farm pollution in 1989 opened up a new area of activity for ADAS by extending its role in advising farmers on waste management, linked to the provision of grant aid for waste facilities. However, the policy shift also inserted the NRA's Inspectors into the regulation of farm waste facilities, giving them a sanction over the approval of grant aid. Not only did this represent an invasion of ADAS's traditional territory, but also made its advisory and grant-approval authority subject to external ratification. How then did ADAS staff deal with these additional responsibilities and challenges?

ADAS supplied technical advice on farm waste management and pollution prevention. Under MAFF's "public good" advisory policy, farmers were offered the opportunity of a free visit from ADAS to provide broad general advice on waste management problems. The advice so dispensed covered the identification of pollution risks and an indication of what types of solution might be appropriate, along with information about the regulations and financial assistance. Typically, during the course of a visit lasting at least an hour, quite complex technical matters would be discussed; but although the advice given might include recommending a particular type of system, free advice always stopped well short of providing an actual design. That could be provided by ADAS's commercial arm, which also offered a service that would draw up detailed plans for waste management systems for individual farmers.

Within ADAS, commercial services were regarded as having a higher priority than statutory work such as free pollution visits (House of Commons Committee of Public Accounts 1991). The availability of free visits and ADAS consultancy services was made known to potential customers through the divisional bulletins sent out to 150 000 farmers and through talks to farming groups and stands at agricultural shows. Alternatively, a farmer might be urged by an NRA Inspector to seek advice on undertaking preventive work or receive a recommendation from a farming colleague who had already had an ADAS visit. Whatever the provenance of the request, though, ADAS visits were entirely voluntary on the part of the farmer. As one senior ADAS official put it emphatically: "We go on to the farm at the farmer's request and only at the farmer's request".

The actions of ADAS field staff were conditioned in three distinct ways: by organizational requirements, by their professional allegiances and by their orientation to their farmer clients. As members of a large bureaucratic and hierarchical organization, their effectiveness and efficiency were typically measured in quantitative terms, and displayed as performance measures and targets achieved. Numbers of free visits made, response times, types of pollution hazard discussed, numbers of consultancy jobs completed, amounts of grant given – measures of these sorts of things – were what counted. Files had to be kept on every farmer visited, and reports compiled and sent after each free visit and consultancy job. Besides the need to report back to clients, checking up was as much to do with accounting for taxpayers' money as anything else.

While the organization broadly determined what the adviser spent his time doing, its content was much less governed. Here, professional rather than organizational allegiances tempered ways of doing things. ADAS is a discipline-based organization. In 1991, approximately 300 professional staff were employed throughout the South West region, covering every discipline involved in agriculture from entomology to economics. The majority of these worked as "front-line" advisers. Since the introduction of charging, advisers had specialized; the traditional generalist, who had his own patch and advised all the farmers in it about anything and everything, had given way to the consultant, specializing in, for example, farm business management, dairy farming, crop protection or pollution control. These consultants were supported by discipline-based regional specialists, including experts in soil science, grassland management, dairy breeding and animal nutrition, who were there to provide specialized back-up expertise.

Farm pollution advice and consultancy services in Devon and Cornwall were delivered by a seven-member, multidisciplinary Farm Waste Team. Two of them provided free advice, normally visiting between six and ten farms a week; the others spent most of their time designing pollution control systems. Pollution control had little tradition as a separate discipline and was not really regarded as such within ADAS. Rather, it had become an extension of other disciplines, such as soil and water engineers, whose main line of work used to be land drainage, and mechanization specialists, who had always dealt with pumps, tanks, lagoons and the like. With pollution control not being an established specialization, there was less scope for field-level advisers to consult upwards within the organization, although there was a national Farm Waste Unit. More than usually, advisers relied on their own technical judgement, informed by their professional outlook as engineers.

For example, the selection of the return period for dirty water systems (i.e. the probable frequency of the scale of storm that would overwhelm a system) was left to the discretion of advisers who based their calculations on a variety of rules. One explained how he had made his decision: "I was asked to do a job. When I asked for criteria by which to do it, no-one could give me any that would satisfy me. That said, I didn't look that far at the time because my senior manager didn't think this was a job I should be worrying about." The adviser settled on one in five years because this "has always been our engineering standard in my mind". He decided it was also "realistic from an environmental point of view. If you start demanding higher standards, that's overcooking." When the one-in-five-year storm did occur, he reasoned, any overflow of dirty water would in any case be very dilute. Farmers often made a similar point – that heavy storms helped cleanse the pollution they caused.

Another adviser remarked that, although the farm buildings experts said that overflow pipes were prohibited, because then a reception tank would not be impermeable as required by the Regulations, he thought that was "nonsense". If there was too much effluent, it was surely better for it to be discharged sensibly than for it to overwhelm the tank. He was "in the middle" and he had to make up his own mind. In the field "you play things by ear".

The discretion assumed by ADAS advisers was within the tradition of their organization as an extension agency, in which field staff were expected to exercise their professional judgement in tailoring available sources of expertise to the specific needs and circumstances of the individual farmer. However, until the mid-1980s, such discretion was exercised within a well established policy framework in which the overall objective – of increasing output and raising productivity – was clearly acknowledged. Since then, the purpose of agricultural policy had become less clear.

Pollution advice was one of the responsibilities that had risen to prominence subsequently, but its precise aims remained unspecified. A pollution control role, in particular, was disavowed. In such a context, where procedural targets (such as the number of free advice visits) took the place of substantive targets (such as reduced levels of pollution), advisers were left with an unusual degree of discretion regarding not only the technical content, but also the specific purpose of the advice they gave.

Advisers involved in farm pollution work admitted that they used to view farm wastes rather differently. Managing animal wastes was always a part of livestock farming, and increasingly farmers had needed advice on storing and disposing of ever larger volumes, but until the mid-1980s the pollution side tended to be downplayed. Some of the advisory staff in the South West had taken part in joint visits to farms with Pollution Inspectors as part of the Farm Campaign begun in 1984 by the old regional water authority. From this experience they had learned not only about spotting the signs of pollution (e.g. how to recognize some of the invertebrates whose presence signified clean water) but also about identifying how it could occur. This had given them a greater appreciation of contamination, not just from large spills of neat slurry, but also from minor leaks and from dirty water.

The third factor influencing the way advisers acted was the need to maintain the trust and respect of the farming community. An ADAS adviser's work was as much about building up relationships and maintaining a reputation as a knowledgeable and helpful adviser as it was about doing the job itself. In such a closed community the grapevine worked fast and farmers soon learned on whose advice they could generally depend. But to satisfy individual farmers it was important that the advice was tailored to their specific circumstances and concerns. Farmers would not follow advice if they perceived it to be impractical. ADAS advisers saw themselves as serving the farmers' interests, and how they interpreted those interests was, therefore, crucial.

In seeing farmers, first, as "our clients", ADAS staff assigned certain interests to them. In particular, the farmer's commercial interests were viewed as predominant, for example in assessing what the farmer "can realistically do". Farmers were also seen to desire "peace of mind", and it was in this context that advisers discussed engineering controls and the need for regular maintenance of systems. Farmers were also seen to need reassurance, and advisers would patiently spend much of their time on visits discussing the implications of choosing one sort of system rather than another.

Secondly, the farmer was a manager. In general, farmers were seen as good managers of animals but poor managers of technology. This was evident both from the way advisers talked about farmers and from the way they negotiated with them during visits. "A lot of farmers think that when they put in a control system that's the end of their problem," one adviser said to us. Another commented that the awkward maintenance jobs were always left until tomorrow: "if something's a lot of hassle it won't get done". In short, "there's always a time when a farmer will be doing something daft".

ADAS staff presented their role in pollution control as helping farmers to avoid causing a pollution incident. The key to that was the installation and management of appropriate facilities for effluent storage and disposal. As one adviser remarked: "our goal is to reduce pollution, and you can only do that by design". The farm pollution problem was not seen first and foremost as one of too much polluting effluent to be managed. Instead, design took as a given that a particular quantity of waste was generated, and sought to solve the effluent containment and disposal problem.

The main obstacle was seen to lie with the financial circumstances of agriculture in general and of individual farmers in particular. But at the same time inaction might also be costly if it resulted in a fine. In other words, the farm pollution problem was seen to pose a financial problem to the farmer. Indeed, it was often presented as just that in discussions with farmers. Pollution was characterized as something that "you can be fined up to £20 000 for". Alongside a sensitivity towards a farmer's financial circumstances, cost-effectiveness was a paramount consideration in the advice that ADAS gave. There was a need, as one of them put it, to find "a solution acceptable to the farmer and his bank balance".

Given that the overall objective was to avoid pollution incidents, cost-effectiveness was not necessarily a matter of looking for a quick and cheap solution, but of identifying when the farmer could afford to implement an appropriate solution. Where there was no immediate pollution problem, ADAS advisers were quite prepared to countenance the continued use of substandard facilities, although they usually warned the farmer that there should not be any difficulty unless a pollution incident occurred. Where there was no obvious cure for a chronic problem, advisers encouraged whatever practical measures might be taken to mitigate it.

In some cases, however, it was quite clear to ADAS advisers that what had to be done was well beyond the financial means of the farmer. As one adviser explained, "you know what needs to be done and you know roughly the income of the farmer and the two don't match". Such farmers were "in a cleft stick": often the farm was simply "in the wrong place", the land was poor, and there had been little investment. They had to be warned of the risk they ran and the prohibitive cost of putting things right, running up to tens of thousands of pounds. The only practical possibility to put to them might be a change of enterprise, say to less profitable beef-cattle or sheep.

The adviser's view of the farm pollution process thus associated not only effluent storage and safe disposal as solutions, but also pollution as an "event" or prosecutable incident, and money as profits, fines and investment costs. In the main, it came down to economics for most farmers. Therefore, it was not a problem "unless you cause a pollution event" (as an ADAS adviser explained to us). Implicit in this comment, of course, was endorsement of the view that causing pollution was rule-breaking rather than a criminal act.

However, providing practical technical solutions to farmers' waste disposal problems was not the only function that ADAS pollution advisers performed on their farm visits. In general, the farmer who requested a visit, already had a reasonable idea about what needed to be done. Although much time could be spent clarifying technical matters – ranging from the general advantages and disadvantages of particular systems down to the minutiae of, say, pump design and operation – advisory discussions also acted as a means for passing on interpretations of the rules and regulations, with the adviser acting as a translator of possible consequences.

The NRA's requirements were thus routinely explained, including the need for its "approval", if grant was going to be claimed, and selected features of the Farm Waste Regulations. Typically, this covered such things as storage requirements, exempt structures and farm waste management plans. ADAS staff translated the NRA's requirements, saying, for example, "that should satisfy them" to one farmer's disposal plan for dirty water, and to another farmer who wanted a pit and a chopper pump (macerator) that "the NRA won't like it".

In many cases, what the adviser was providing was reassurance (something that farmers expected experts to provide; see Ch. 6). Typically, the farmer had already decided what was to be done and was able, financially, to do it. The adviser essentially confirmed the appropriateness of what the farmer proposed and might influence minor details. But in other cases, what was being sought, and was sometimes provided, was implicit or covert sanctioning of rule-breaking. Typically, the farmer had a good idea of what ought to be done, but was unable or unwilling to do it, or all of it. The adviser judged what the farmer could *realistically do*, and the result was often advice of the form "you needn't do it yet".

With reference to a substandard silage effluent control system, one adviser warned the farmer "It can continue to be used until you cause some pollution". On another farm, an earth-banked slurry pit would not meet current requirements, but the adviser said "Frankly, I'd leave that one for now", and advised the farmer to concentrate on his more pressing dirty water problems.

In several cases it seemed clear that the farmer intended to do "a DIY job" that might repair the worst of the problem for a while. It was against the Regulations for farmers *not* to tell the NRA, but, as long as they did not claim grant, they were unlikely to be found out. Such solutions were cheaper for the farmer. Advisers did not rule them out. "Better to do something than nothing", they reasoned.

Advisers even turned a blind eye to cases where it was likely that a farmer would do what the Regulations specifically prohibited. One farmer had built a silo in 1990 and, to save money, wanted to pipe the silage effluent into the new slurry pit for which he was seeking grant aid. The Regulations, however, specified that silage effluent receptors should be impermeable for 20 years without maintenance and it would be very expensive to construct a slurry pit to that design standard. So, the adviser suggested installing a separate silage effluent tank at a cost of about £2500. "That's the party line" he commented later, "What he will do in practice, I'm sure, is to put that pipeline in [to the reception pit] and say nothing about it in his grant claim!"

ADAS advisers specifically disavowed a role as pollution regulators. As one of them commented "virtually every farm I go to is polluting to some extent or another, but it's not my job to tell them so". They saw their task instead as suggesting remedies for deficiencies in waste management and disposal practices. Of course, they would point out to a farmer if he had a problem that was likely to get him into trouble. But they did not take the matter any further and certainly did not see it as part of their responsibility to report the pollution they came across during their farm visits. That would have been a breach of their "client's trust", and counterproductive if farmers became wary of consulting them. One adviser commented that on only two occasions, both involving obstreperous farmers and serious pollution, had he notified the NRA about something he had seen on a farm visit without first getting the farmer's permission. He had done so surreptitiously.

Much of the interaction between ADAS and NRA field staff was actually mediated through the farmer, with ADAS officers seeing their main role as steering farmers through the NRA's procedures. ADAS tended to talk about the NRA in terms of its official roles and how its staff fulfilled these: they detected water pollution, informed farmers if they were causing it, meted out punishment to the perpetrators of pollution and "approved" designs for new pollution-control facilities. With the exception of the last, these roles were accepted and their performance generally, if sometimes begrudgingly, was respected. In the words of one adviser: "They are very good at inspecting watercourses . . ." This sort of remark put the Pollution Inspectors in their place, at the watery margins of the farm. Resonating with more overtly critical comments by farmers, such as that "they care more about fish than farmers", it implicitly asserted ADAS's stance as the only people who understood agriculture.

In contrast, ADAS staff strongly and explicitly contested the NRA's right to the role of "giving the go-ahead" for a farmer to proceed with the construction of an effluent installation on which grant would be claimed. To continue with the words of the adviser quoted above: ". . . but not very good when it comes to engineering . . . The NRA set themselves up as judge and jury when they don't know anything", and to quote another: "In my mind they cannot reject any [proposals for waste facilities] because they don't have the expertise. They're not qualified to judge that sort of thing". At the same time, though, ADAS staff argued that the NRA should provide design guidelines.

This is not the contradiction it may seem. While ADAS contested whether NRA staff were qualified to judge the engineering aspects of design – design loads, construction standards and so on – the guidance they sought concerned the level of pollution risk incorporated in the design. This risk was embodied in the "return period". Implicit in this notion was that pollution would inevitably and periodically occur. What ADAS staff regarded as a stark engineering reality meant that selection of any practical capacity standard to contain pollution was, at one and the same time, acceptance that pollution would occur in the extreme circumstances when the standard was exceeded. The risk could be reduced, but not eliminated, and only at extra cost by increasing the storage capacity. Rhetorical references to the need for NRA guidance on such matters were actually, therefore, to be more of a challenge to absolute notions of pollution control and an assertion that the real choice, in which the NRA was irredeemably implicated, was between more or less pollution at less or more cost.

In practice, ADAS's designs were rarely rejected by the NRA. According to one ADAS adviser, this was because the Pollution Inspectors "rely on the designer to get it right", implying that the NRA acquiesced in ADAS's self-definition as the engineering experts. Thus, on the ADAS frontline the NRA were pollution experts but not engineering experts and were certainly not experts in agriculture. Significantly, the rural Pollution Inspectors acquiesced in these characterizations, even though some of them did have agricultural training and experience, and all of them had an appreciation of farming issues. As one of them remarked: "We are careful not to tell farmers how to run their farms . . . We don't give advice except on pollution risks. We avoid treading on ADAS's toes."

Local environmentalists and farm pollution: the moral discourse

Whereas ADAS staff reinforced the notion of farm pollution as a technical problem and a form of rule-breaking, the NRA's Pollution Inspectors had come to embody, for farmers, a new morality in the countryside which saw pollution as something discreditable – a form of environmental crime. As such, Pollution Inspectors were indicative of a changing rural society in which agriculture and its prerogatives were increasingly under challenge. New people were living in the countryside with ideas about how it should be managed which often differed sharply from those of traditional production interests.

Counter-urbanization had been a feature of most of rural England since the 1960s, turning it into a predominantly middle-class territory. Devon's population growth had been among the highest, totalling 13 per cent between 1970 and 1990 (Champion & Townsend 1990: 54). In contrast to, say, the South East and the Midlands, the bulk of those moving into the rural areas of the South West were not commuters but long-distance migrants and retirees, many of them attracted by the high quality of the environment (Bolton & Chalkley 1990). By 1991 four out of ten households in East Devon were occupied solely by pensioners. During the 1980s, the migration flows that previously had impacted largely on the coastal and market towns also began to affect the villages and the deep countryside. The completion of the M5 motorway opened up many parts of rural Devon to long-distance commuting, but residential dispersal was also encouraged by the liberalization of the planning system and facilitated by farmers themselves, many of whom (with government encouragement) sought to realize some of their assets by converting redundant farm buildings into expensive dwellings (see DOE/Welsh Office 1992, Kneale et al. 1992). The social fabric of once small agricultural villages and hamlets was transformed. The influx of large numbers of newcomers was also associated with, and helped catalyze, a major shift in public attitudes to agriculture and the countryside. Many farmers had new neighbours with quite different perceptions of the function of the countryside. Of the dairy farmers surveyed in Devon, one in six had experienced direct pressure from neighbours and local people to change their farming practices.

The scale of regional growth and the social change associated with it led to increasing prominence in local politics for environmental issues. The combination of an attractive countryside and in-migration meant that local amenity groups flourished. The Green Party recorded its greatest achievements in the region, where at times it seemed on the verge of a major breakthrough, most notably in the 1989 European elections when it received over 20 per cent of the popular vote. Amid this rising environmental consciousness, a local network of newspapers, environmental groups and activists ensured a high profile for farm pollution issues, which kept up the pressure on the NRA. The regional newspaper, the *Western Morning News*, played a key role, and its coverage of the findings of South West Water Authority's (1986) report on pollution in the River Torridge helped make farm pollution a public issue in the region. Subsequently, farming leaders, regulators and advisers have had to deal with repeated press enquiries.

Friends of the Earth (FOE) were the group most active nationally and locally in campaigning on the issue of farm pollution. As part of the rising level of environmental concern and activism in the late 1980s, the local FOE groups in Exeter and Plymouth had seen their memberships grow and new local groups had been formed in North Devon, and in Honiton and Sidmouth in East Devon. These local groups were especially active in fundraising, and it had been agreed with FOE headquarters in London that some of the money raised in the South West should be returned to the region to fund a particular local campaign. The majority opinion among local activists was for a campaign against farm pollution, which seemed to be such a growing problem in the region. In the run up to the privatization of the regional water authorities and the formation of the NRA, water pollution had become a high-profile issue nationally in which FOE had become prominent. In its evidence to the House of Commons Environment Committee's investigation into Britain's deteriorating rivers, FOE had drawn heavily on the Torridge Report (see Ch. 4) to identify farm effluent as a growing threat to the water environment that demanded stronger regulation. FOE headquarters wanted to get a better purchase on the issue of farm pollution and so it agreed in 1990 to finance a campaign in the South West, supporting one full-time campaigner in Devon and one in Cornwall for a six-month period. The main aims were to draw attention to the problem, bring pressure to bear on the NRA to be tougher in prosecuting offences and to draw up a manual to assist local groups elsewhere to combat farm pollution.

The local campaigners were to work under the broad direction of FOE's Water and Toxics Department. This department was one of the strongest in FOE, previously having been run by the much-respected Andrew Lees, who had conducted successful campaigns to draw attention to the issue of pesticide contamination of drinking-water supplies in the late 1980s. By 1989 Lees had gone on to become FOE's campaign director.

The work in the South West was to feed into and inform FOE's national activities on water pollution. As with its other campaigns, this had a strong orientation towards the protection of nature as a vital but besieged force, expressed in FOE's slogan "Help the Earth to fight back". The group wanted to see water-quality policy and monitoring reformulated on an ecological basis, with natural objectives and criteria to replace the existing ones based on human uses of water. This shift from a relative to an absolute definition of environmental protection, with its orientation to "the ecological integrity" or "the ecological health of watercourses" (House of Commons Environment Committee 1987: 160, 162), was argued for by Andrew Lees in the following terms:

> a duty . . . to take action to prevent injurious pollution occurring to flora and fauna must lie at the heart of any policy for protecting the water environment. If we had in our rivers the flora and fauna consistent with the ecological potential of the rivers we would then know that . . . the rivers were healthy and also suitable for other uses. The biota of a river is its most sensitive characteristic; the diversity of biota is a reassurance that river quality is high. (ibid.: 214)

From such a perspective, a natural river was not just a resource but an ideal, an exemplar and a guarantor. It provided an absolute standard against which to judge the deterioration of other rivers and to gauge any progress made in their improvement. The protection of the best quality rivers was therefore an overriding consideration, and FOE was dismissive of overall assessments of trends in water quality that offset improvements in rivers of poorer quality against deteriorations in rivers of better quality. FOE saw the greatest threat to the best quality rivers coming from agricultural pollution. In the words of one of its campaign guides, "farm pollution strikes at some of our most environmentally valuable watercourses" (FOE 1991). The threat, moreover, was both stark and insidious: "Sometimes the damage is obvious – large discharges mean dead fish. Even small seepages can cause a subtle reduction in the diversity of aquatic flora and fauna" (ibid.: 160).

FOE's national staff had had little direct experience of farm pollution, but they believed that it was seriously under-recorded because of the lack of monitoring of rural catchments, and as a result was not treated with the gravity it deserved. There was a desire, therefore, to demonstrate not only that there was more pollution going on but also that the problem was much more grave than was officially acknowledged. This required FOE to become involved in the detection and regulation of water pollution, which was the task assigned to the two farm pollution campaigners in the South West. It wanted thereby to be able to influence the way in which the newly created NRA fulfilled its responsibilities and to demonstrate the need for many more resources to be devoted to monitoring and enforcement. To do so it had to politicize pollution regulation.

The campaign in Devon and Cornwall thus had several components. One was to penetrate the NRA's regulation of farm pollution incidents to ensure that it treated each with due dispatch and rigour. A former water authority pollution officer acted as an adviser to FOE on the campaign. He drew up a manual on how to spot and record farm pollution and when were the best times to look for it during the farming year. Armed with this information and equipment to take water samples, the two FOE farm pollution campaigners began patrolling the rivers of the South West looking for farm pollution. They started at the beginning of March when they had been advised that "slurry stores may be full and overflowing" (FOE 1991). Whenever they encountered any pollution, they alerted the NRA, and during the first month 15 farms were reported for polluting (*Western Morning News*, 14 June 1990). The time taken for an Inspector to arrive on the scene would be noted; if there was a significant delay a complaint would be lodged. One of the campaigners complained to the *Western*

FARM POLLUTION
IS
KILLING RIVERS!

IF YOU SEE RIVER POLLUTION CALL THE NATIONAL RIVERS AUTHORITY (NRA) ON

0800 378500

AND
CALL FRIENDS OF THE EARTH'S FARM POLLUTION MONITORING UNIT ON

0392 423308

Friends of the Earth

Figure 7.1 "FARM POLLUTION IS KILLING RIVERS!". (Source: Friends of the Earth)

Morning News of delays of up to 14 days, "by which time all the evidence had disappeared" (ibid.: 5 April 1990). Response times generally were collated and reported regularly to the NRA's regional headquarters. Not surprisingly, it was found, "they pulled out the stops more and more".[1] The FOE campaigners would then observe the Inspectors' handling of the incident. If they had already taken formal water samples themselves, they would inform the Inspector, thereby obliging him not only to do the same but also to treat the incident seriously. They found that "Formal samples can be used as an instrument to persuade the NRA to take decisive action" (FOE 1991), and they warned publicly that "If the NRA is unable to collect evidence, as a result of understaffing, the FOE is prepared to take formal samples and follow through with legal proceedings" (ibid.). Afterwards, the campaigners would follow up, with phone calls and letters to the NRA, any incident they had reported, checking up on progress and complaining of any delay or apparent laxity in dealing with the matter.

FOE was critical of the small proportion of incidents that were actually prosecuted. According to the Devon campaigner, "the Inspectors on the ground were pretty dedicated to preventing farm pollution, but action would get stopped by those above them." Too much emphasis was being placed by the NRA on the "together we can beat it" approach – "too much of the carrot and not enough stick". To keep up the pressure, the campaigners would alert not only the NRA but also the media to the more severe incidents they encountered. They were able to attract coverage in the regional newspaper, the *Western Morning News*, and on regional television. The Devon campaigner had a regular column in a local free newspaper, the *Exmouth Herald*. The campaigners distributed leaflets to local FOE members, and posters to libraries and village noticeboards, asking people who spotted farm pollution to telephone the NRA and FOE (Fig. 7.1). The image of West Country farming, which FOE had to overcome, was typically a cosy one of farm families looking after their cows and their pastures, producing wholesome milk and purveying cream teas. FOE's publicity emphasized instead the menace of farm pollution and the threat it posed to the water environment, pointedly passing over its social origins. FOE's widely circulated poster declared starkly "FARM POLLUTION IS KILLING RIVERS" – a message whose explicit animism emotively encapsulated the new environmental morality.

Seen as environmental crime, pollution should be prosecuted: a successful prosecution served to confirm wrongdoing. FOE's line was that all instances should be prosecuted where the evidence justified bringing a case

(House of Commons Environment Committee 1987: 162). A prosecution also raised the profile of the issue: it attracted publicity, which in turn broadcast the moral message. In the words of the Devon campaigner, "It would only be seeing a lot of farmers in court that would bring the pressure to bear on the farming community". The campaigners, though, found themselves in a conundrum: often it seemed that, without publicity over a pollution incident, there was not so much pressure on the NRA to make an example of the farmer, but that the media was less interested if a prosecution was not in sight. It was vital for the campaigners to lift farm pollution out of the routines of farming practice and the recesses of regulatory bureaucracy, and to give it a public profile as a shocking issue. The strategy to achieve this was to target prominent farmers and "blow the whistle on them." FOE groups were asked to "look out for your local big farmer – see if he's polluting".

The strategy was influenced by a fear that the media might be becoming inured to farm pollution incidents: "There were so many that it had become a bit run of the mill", the Devon campaigner told us. Targeting the larger and more prominent farmers also avoided the risk of adverse publicity from being seen to get "some small poor farmer" into trouble. The *Western Morning News* was reluctant to run stories that seemed to be anti-farming. The campaigners also feared that the NRA might be able to dismiss cases they brought to light which involved more vulnerable members of the local farming community. A prominent and wealthy farmer, however, would bring more adverse public attention. "People would say what the hell are the NRA doing" and this would help "buck the NRA into action". One notable "scalp" was the then vice-chairman of the county branch of the NFU (National Farmers' Union). The Devon campaigner, alerted by a local angler informant, came across pollution coming from the farm in the Torridge catchment, took samples, and called the NRA and the BBC. The farmer was eventually fined £1240.

Environmental activists were not the only ones vigilant against farm pollution. Riparian owners – particularly those from outside the area and not themselves engaged in farming – had become more energetic in pursuing pollution issues and protecting their fishing interests. Their efforts were stimulated not least by the powers and consultative structures of the new NRA. Anglers were active in reporting incidents. Local amenity groups also undertook organized river walks, partly to "keep an eye out for pollution".

The FOE campaign in the South West had several results. First, it demonstrated the under-reporting of farm pollution incidents. The South West had the worst record of farm pollution incidents in the country, with 589 reported in 1989. However, FOE alarmingly suggested that "the actual figure could be more than 10 000" (*Western Morning News*, 5 April 1990). The Devon campaigner, writing to the *Western Morning News* about the number of incidents reported by FOE ("despite the fact that FOE is made up of ordinary members of the public who . . . have no experience in detecting farm pollution") concluded: "Clearly, therefore, farm pollution is a much more serious problem than the farm industry would like us to believe. It also shows that the NRA is hopelessly understaffed" (*Western Morning News*, 14 June 1990).

Secondly, the campaign did reveal that the NRA staff resources given over to farm pollution monitoring and enforcement were a limiting factor. FOE made much of this in their publicity. It was clearly an objective of their campaign but also part of a strategy to encourage the NRA to see FOE as an ally (significantly, FOE studiously refrained from publicly criticizing the NRA, except for being insufficiently resourced and staffed). As the Cornwall campaigner complained "unless they [the NRA] are given more manpower, farmers will continue to get off the hook at the expense of river quality" (*Western Morning News*, 5 April 1990). Writing in the NRA's own newspaper, FOE's national water campaigner reasoned that, with less than one Pollution Inspector for every thousand dairy farms in the South West region, "the level of enforcement is almost not enforcement. It is a fire brigade service" (quoted in *Western Morning News*, 2 August 1990). These criticisms did have some effect on the NRA (see below).

If the NRA was seen essentially as an ally, it is clear where FOE pinned the blame for farm pollution. The third result of the South West campaign was a *Campaigners' guide to farm pollution* which distilled the lessons learned and made them available for local groups and activists nationwide (FOE 1991). The guide did not mince its words in identifying "THE CULPRIT – MODERN AGRICULTURE" adding that "Today every modern farm must be regarded as a potential threat to the environment." The guide proposed that "Identifying and reporting farm pollution is a way for everyone to do something for the local environment", and in simple terms it described what to look out for, when and where farm pollution might occur, and what steps to take if it were found. "Shopping" your local farmer was what responsible citizens should do.

The NRA, Pollution Inspectors and farm pollution: negotiating the moral discourse

Although farmers and the pollution they caused were the object of FOE's campaign, its objective was to steel the NRA to take decisive action. The agency in turn sought to identify with the environmental movement and draw on the movement's moral authority, but it was subject to other pressures and constraints, which meant that its approach to farm pollution in practice was not so single minded or straightforward. The NRA needed to be seen to be firm in its handling of polluters, but also to be reasonable and even-handed in dealing with different industrial sectors, including agriculture, to ensure their cooperation in improving water quality. This led to efforts to develop a more consistent, defensible and high-profile approach to prosecution, to be pursued alongside the more traditional means of pollution control, namely those of persuasion and preventive regulation. The agency's field staff had the task of reconciling these contrasting approaches. In tackling farm pollution, they operated not only within a framework of NRA policies and procedures, but also through their own local networks of colleagues, informants and farmers. They thus had to negotiate conflicting discourses about the rights and wrongs of pollution.

Through its publicity and consultations, the NRA aligned itself with popular concern for the environment. Moderate conservationists were brought onto its board and its Regional Rivers Advisory Committees, alongside the more traditional industrial and agricultural representatives. The Authority committed itself to undertake its statutory liaison duties through "a much more open consultation process with . . . conservation bodies than pertained under previous arrangements" (NRA 1989a). The NRA also paid considerable attention and resources to publicizing its work. It claimed as a key achievement of its first year of operation the establishment of "a corporate identity and strong public image . . . through effective media relations" (NRA 1990a: 4). Its eye-catching logo of an arching salmon was displayed prominently on its vans, buildings, leaflets and posters. Press releases which concentrated on "publicising pollution incidents and their effects, and the NRA's regulatory role" (ibid.: 39), projected the image of the authority as an environmental guardian and pollution watchdog. A key means of actively associating the public with its work was through the reporting of pollution incidents. Following the FOE campaign, the NRA South West had set up its 24-hour free pollution hotline. The encouragement of people to be vigilant over pollution gave it the air of a moral crusade.

Notices alerting the public to the threat to rivers from farm pollution, and encouraging people to report any incidents, were visible on most village notice boards and pinned to telegraph poles in even the most isolated parts of the Devon countryside. Hotline cards were to be found in local libraries, police stations and even the local branch office of the NFU. In general, its public relations drew upon and encouraged expectations that the NRA would take a tough stance on pollution, and this was indeed the agency's declared intention – to take polluting events seriously, to pursue a higher rate of enforcement and to press for stiffer penalties (NRA 1990a).

The creation of a national body from the separate regional authorities provided the opportunity to develop a strategic outlook on how to protect and improve the rivers of England and Wales, and to build up a central body of scientific and policy expertise. It began to address the criticism made by both environmental and industrial lobbies of varying standards and approaches under the old regional water authorities, by setting out national policy standards and taking steps to ensure their consistent application. In 1990, NRA headquarters started issuing a series of policy implementation guidelines, and in 1991 it created a new post of Director of Operations, with responsibility for overseeing the application of national standards (NRA 1992b: 11).

As part of this more strategic approach to its work, the NRA began to prepare studies of major problems in water quality. One of the first of these was on agricultural pollution, prepared by the NRA's Farm Waste Group (the successor to the WAA's Farm Waste Group (see Ch. 4) and including many of the same members). However, the consultations surrounding its preparation brought to a head the antagonism of farming organizations. What was seen as a confrontational stance by the NRA was causing growing alarm among farmers. There was a strong feeling that they were being unfairly targeted by the new regulators. The BBC radio programme, *Farming Today*, reported that "Farmers do feel that the NRA is gunning after them", to which an NRA spokesman responded "We try to be evenhanded between farmers and any other polluters" (*Farming Today*, 11 May 1990). Of even greater concern to farming leaders was the NRA's challenges to production and policy decisions. The chairman of the NRA had declared publicly that "in many areas of intensive livestock production, present levels of stocking and slurry disposal are damaging the environment" (*Financial Times*, 31 October 1990). Within the NRA there was a strong feeling that the farm waste problem was endemic in certain regions (including the South West) with many of the dairy farms overstocked, in the wrong place or having production systems that were intrinsically risky. Fundamentally, it was felt that the fault lay with agricultural policy for having encouraged the intensification of dairy farming. From this perspective, the prosecution of individual farmers was dealing with the symptoms of the problem. It was much more important to seek the redirection of farming practices and policies towards ones that posed reduced environmental risks and pressures.

Whereas the regional water authorities had had no standing in agricultural policy circles, the farm pollution issue had pitched the NRA into a significant position. It was closely consulted on the Farm Waste Regulations, which were drawn up by its main parent department, the DOE. The 1989 Water Act specified the NRA as a statutory consultee in the drawing up of the *Code of good agricultural practice*, and the NRA was not only closely involved in the preparation of the code by MAFF but also was represented on MAFF's Farm Waste Committee. Although NRA officials had trenchant criticisms of agricultural policy, they had to be careful not to overstep the bounds of their acknowledged expertise. They were pollution specialists, not agricultural specialists (which was the role of ADAS). What they sought to do therefore was to draw the environmental constraints of waste management and disposal more centrally into production decision-making. They did this through the notion of farm waste management planning.

The idea of a farm waste management plan was a simple one of identifying, on a map of the farm, the areas suitable for waste disposal: taking account of topography, geology and soil type, the location of drains and watercourses, the season and rainfall. For the NRA, though, it had a much larger purpose. It built upon the NRA's wish to get away from the technical fix mentality that grant aid and the Farm Waste Regulations had encouraged, and to pursue a didactic approach instead. In the words of a member of the Farm Waste Group: "Farmers understand their land more than they understand a new-fangled storage system, so you should get them to concentrate on that; this is what the farm waste management plans do". The intention was "to get farmers thinking a little like waste treatment engineers". The systematic preparation of farm waste management plans would also begin to indicate which farms were overstocked and which ones were ill sited. Such plans were thus seen as the key components of a strategy of improved planning for the disposal of farm wastes. Individual plans would be prepared and implemented in relation to the needs of river catchments, which was the preferred planning framework for water professionals (see Ch. 4). The completion of plans for a catchment would help pinpoint the risks to the river system and provide the basis for judging whether the catchment was overstocked. The chairman of the NRA's Farm Waste Group called for all farms to have a waste management plan, approved by the NRA; those without sufficient land capacity to absorb the waste produced would have to reduce their herds (ENDS 1990). The consequence would have been to draw this aspect of farm management, and those who advised on it, into the ambit of the NRA's catchment planning approach – an approach premised on the proposition that the NRA had to influence land use in order to protect water.

The NRA pursued the notion at successive stages in the development of farm waste policy. For example, where a farmer sought exemption under the Farm Waste Regulations from the requirement for at least four months slurry storage capacity, the NRA required the preparation of a farm waste management plan to demonstrate that the slurry could be safely spread throughout the year. The NRA also introduced into the *Code of good agricultural practice* the basic steps of survey and assessment that together would comprise a simple practical waste management plan.

The NRA's major report on agricultural pollution was to be the vehicle for setting out its strategy on farm waste management plans. During the preparation of this report, however, the NRA came under strong pressure from agricultural interests, both behind the scenes and in the press. The NFU felt that the powers already given to the NRA under the Farm Waste Regulations were "arbitrary and even excessive" (NFU 1990: 2). The NFU president wrote to the chairman of the NRA, challenging its objectivity in preparing the report (*Farmers Weekly*, 18 October 1991). The farmer members of the NRA Board (nominated by MAFF and the Welsh Office) also objected strongly to the draft report and there was unease within both MAFF and ADAS. According to the report's chief author, "we had our knuckles rapped for straying too far into land use policy".

When the report eventually appeared in January 1992, its tone was strikingly conciliatory (NRA 1992a). In the preface, the NRA's Chief Scientist, Dr Jan Pentreath, reasoned that the prominence of "the agricultural industry" as a source of pollution was "not because it is less controlled or more careless than any other, but because of its immense size, the complexity of the tasks which it undertakes, and its virtual ubiquitous presence in all catchment areas" (ibid.: 3). This was seen to justify an industry study of this type, but further reassurances were given that farming was not being singled out for a more stringent approach. "The NRA", Pentreath reassured, "is committed to working together with such industries, particularly the farming industry with which such good relationships have been built up in the past". While acknowledging that the Authority "must, however, remain a firm regulator if environmental improvements are to be attained", he ruled out the need for any additional legislation and looked to the production of "reports such as this", to clarify the problems "such that both the NRA and the farming industry can work together to achieve the improvements which are sought" (ibid.: 3).

The report reviewed the causes, scale and impact of different types of agricultural pollution, and set out the NRA's strategy for dealing with them, centred upon a "catchment management planning" approach within which pollution risks from individual farms could be addressed by means of waste management plans. The NRA called for grant aid to assist farmers in drawing up these plans, and for the adoption of such plans to be a prerequisite for grant support for other pollution prevention measures.

On the same day that the NRA's report was published, MAFF announced the establishment of a pilot scheme to address farm waste pollution through farm waste management plans (MAFF 1992). Although what was proposed fell short of the NRA's ideal – plans were to be voluntary, not grant aidable, and not necessarily with any expert input – it marked the accommodation of MAFF's and the NRA's positions. From this point onwards, the NRA sought to depoliticize the farm waste problem.

The NRA ceased to issue annual reports on farm pollution incidents. The first of these, back in 1986, had established agricultural pollution as a public problem (see Ch. 4). It was now deemed that singling out agricultural problems was neither "equitable, nor a sensible basis upon which to proceed" (NRA 1992c: preface) and they were replaced by reports of water pollution incidents from all sources. The role of prosecutions was also de-emphasized. The NRA's major report on agricultural pollution of 1992 had made no reference to prosecution policy, and in 1993 the NRA made it clear that "the approaches required for preventing pollution and dealing with agricultural waste are much broader than legal action", and embraced "catchment management planning, waste disposal planning, farm waste management plans, farm visits and cooperation and liaison with the farming community" (NRA 1993: 47).

Despite the earlier rhetoric of the NRA, it is evident that there was greater continuity than change in the way in which industry in general and agriculture in particular were dealt with. There was still considerable reliance on the sort of information campaigns and persuasion that had been pursued by the former regional water authorities. Although the NRA's tough words might have alarmed farmers, the reality "on the ground" was different, as we have seen in Chapter 5. However, as it edged away from confrontation with the farming lobby, the NRA had to face a sceptical environmental lobby.

The contradiction between the NRA's rhetoric and reality emerged most starkly in relation to the pollution incident statistics. By encouraging the public to report incidents, the authority had let loose a genie. With increased vigilance from the public and environmental activists, the Authority was faced with a rising level of reported pollution incidents (Table 7.1). However,

Table 7.1 Farm pollution incidents in England and Wales and the South West

	Under RWAS		Under NRA				
	1986	1988	1989	1990	1991	1993	1994
Reported	3427	4140	2889	3147	n/k	n/k	n/k
Substantiated	n/k	n/k	n/k	n/k	2954	2883	3329
Serious	622	940	522	644	n/k	n/k	n/k
Major	n/k	n/k	n/k	239	99	63	36
% of reported/substantiated incidents							
Serious	18.1	22.7	18.1	20.4	n/k	n/k	n/k
Major	n/k	n/k	n/k	7.6	3.3	2.2	1.1

	Under South West Water Authority		Under NRA South West			
	1986	1988	1989	1990	1991	1992
Reported	830	836	589	782	n/k	n/k
Substantiated	n/k	n/k	n/k	n/k	718	686
Serious	74	420	160	173	n/k	n/k
Major	n/k	n/k	n/k	71	25	13
% of reported/substantiated incidents						
Serious	8.9	50.2	27.2	22.1	n/k	n/k
Major	n/k	n/k	n/k	9.1	3.4	1.9

Sources: for Tables 7.1 and 7.2
See Table 7.2.

Table 7.2 Prosecution of farm pollution incidents in England
and Wales and the South West[#]

	Under RWAS				Under NRA			
	1985	1986	1987	1988	1989	1990	1991	1993
No. of prosecutions instigated	159	128	225	148	161	123	159	96
% of reported incidents	4.5	3.7	5.8	3.6	5.6	3.9	5.4*	3.3*
% of serious incidents	28	21	23	16	31	19	n/k	n/k
% of major incidents						51	161	152

	Under South West Water Authority				Under NRA South West		
	1985	1986	1987	1988	1989	1990	1991
No. of prosecutions instigated	23	31	19	22	30	40	26
% of reported incidents	3.7	3.7	3.4	3.2	5.4	5.1	3.3
% of serious incidents	30	42	5	6	20	23	n/k
% of major incidents						56	104

Notes
\# Rates relate to incidents occurring in the stated year for which prosecution
proceedings have been started (whether complete or not). Prosecutions, in
the stated year, of incidents occurring in a previous year are not included.
The "rates" given for serious and major incidents in this table are also
slightly misleading as they relate to the proportion of those categories
which would be covered by the number of prosecutions taken. Thus, there
is no certainty that, for example, 51% of major incidents were prosecuted
nationally in 1990, only that the number of prosecutions would account for
51% of major incidents. More information is available for 1993 when only
21% of major incidents were actually prosecuted, the remainder of the
prosecutions were for significant incidents. The figures, thus, have a tendency
to overestimate the prosecution rates of serious and major incidents.
* Rates given are of substantiated incidents.
Sources: for Tables 7.1 and 7.2
WAA/MAFF 1986, 1987, 1988, 1989; NRA/MAFF 1990; NRA 1992c, 1992d, 1993,
1994, 1995; figures supplied by Pollution Control NRA South West, 1992 and
NRA HQ Bristol.

this was not matched by a significant increase in the number of pro-secutions (Table 7.2), and the paradoxical consequence was a falling rate of prosecutions. This did not impress environmental groups, and the NRA began to attract the type of criticism – of organizational laxness or even complicity – that had tarred the regional water authorities. In 1991 the Commons Public Accounts Committee pronounced that too few polluters were being taken to court and that farmers in particular were not being submitted to the "polluter pays" principle (House of Commons Committee of Public Accounts 1991).

The NRA responded by seeking to establish a more consistent and just-ifiable approach to prosecution. To ensure greater uniformity between and within regions, nationally coordinated legal training was established for all new Pollution Inspectors, covering pollution and case law, report pre-paration and the types of evidence required, together with a mock trial. But the most critical and decisive action of the NRA was to alter the way in which pollution incidents were classified and reported, and at the same time to introduce a set of national prosecution guidelines related to the new classification. These changes, which were introduced in 1991, meant that the Authority could counter criticisms from both the environmental and agricultural lobbies and address its own ambivalence over how pollu-tion and polluters should be treated.

One change was to replace the "reported" incident category (previously taken as the overall level of incidents) by a "substantiated" incident cat-egory, such that incident reports from the public were counted only if pollution had been verified by an Inspector (NRA 1992d). Reported incid-ents that were not verified were excluded from the total. This allowed the NRA both to loosen the dependence of its prosecution rates on increased reporting by the public and to address the NFU's criticism that the reported incident totals relied on public interpretations of pollution – of course, the change made the total dependent on the availability of Pollution Inspec-tors instead.

The other change was to the way incidents were then classified. Previ-ously, they had been divided into "serious" incidents and the rest; the serious category accounted for about 20 per cent of the total. Now there was to be a threefold classification of "minor", "significant" and "major" incidents. The new "major" category was to include "only the most severe incidents" (NRA 1992c: 13).

This change in the classification system was accompanied by the intro-duction of national guidelines, which advised that *all* major incidents be prosecuted if there was "sufficient evidence to take such action" (NRA 1993: 40). By regrouping incidents in this way and pledging to prosecute the most severe, the NRA could publicly present itself as taking a clear and principled stance. It could thus respond to the imperatives of the environmental lobby and counter its criticisms of organizational ineptitude. At the same time it relieved the moral dilemma of Pollution Inspectors by bringing these severe incidents within the realm of environmental morality, where the seriousness of the offence overrode the question of the culpability of the perpetrator. On the other hand, by leaving discretionary the issue of whether or not to prosecute "significant" incidents the NRA acquiesced to the technical discourse of the agriculturalists and assuaged the farming lobby who had accused the NRA of being too ready to prosecute. It also allowed Pollution Inspectors to continue working with farmers with lesser pollution problems, while reserving to them the sanction of prosecution as a threat against particularly troublesome or deviant individuals. The scope retained for discretionary action was considerable – the "significant" category included the large majority of incidents that previously would have been judged "serious" – thus leaving room for the negotiation of pollution control at the farm level.

How the new guidelines operated was obscured by the NRA's reporting of prosecution rates, which did not break them down by incident category, so that it was impossible to judge the NRA's performance. With the level of prosecutions high in relation to the number of major incidents, the impression was given of a firm regulatory stance. Unfortunately, the prosecution figures embraced unrevealed numbers of lesser incidents, and so it was unclear specifically what proportion of major incidents were prosecuted.

Any image of firmness towards agriculture was undermined when incident figures for 1993 were released in a form distinguishing the categories to which prosecutions related. These revealed that only 20.6 per cent of the major farm pollution incidents that occurred in that year were subsequently prosecuted (NRA 1994: 54–5). A subsequent report by the National Audit Office revealed that just 9 per cent of the most serious farm pollution incidents were prosecuted by the NRA in 1993 (National Audit Office 1995: 2). Evidently, the staff on the ground were still struggling with the NRA's ambiguities. The dual message that the Pollution Inspectors received from their own organization – prosecute more but still work with farmers – was compounded by the conflicting pressures they experienced on the ground

– from the public, the press and environmentalists, to be tough; and, from farmers and agricultural advisers, to be conciliatory.

The much publicized moral mandate of the new Authority to act as a "water guardian", together with increasing public concern over the environment, certainly impacted on the Pollution Inspectors. By the early 1990s they had come to regard pollution as an environmental "crime" rather than a breaking of the rules. Certainly, there had been a fall in their threshold of tolerance of pollution from what it had been in the 1970s (see Hawkins 1984, Knowland 1993). While very much sharing an environmentalist perspective on the problem, the Pollution Inspectors did not accept the single-minded approach to prosecution demanded by FOE. Indeed, the FOE farm pollution campaign caused some resentment and irritation. One Pollution Inspector commented that, if FOE find something, they expect action on it immediately. He believed they "don't really understand what's involved" in regulating farm pollution and instead they simplified the issue.

Although acutely aware that effective pollution control involved much more than instigating legal procedures, Pollution Inspectors nonetheless felt there was a moral justification for increasing the number of prosecutions and recognized the benefits of such a strategy in terms of public support. One Pollution Inspector felt that, whereas the public had had "very little faith" in the former water authority, the NRA was more popular because of some "crucial" prosecutions and the winning over of the press. Pollution Inspectors valued public vigilance in their patches. Pressures also came from their bosses in Pollution Control, who told them that it was NRA policy to prosecute more and if in doubt to take a formal sample.

However, Pollution Inspectors had reservations about how far this more stringent approach should, or could, go. One set of doubts related to the judicial process. Although Inspectors acknowledged that a serious pollution incident deserved punishment, the fines handed down were not felt to fit the crime; many were considered "derisory". Despite the raising of the maximum fine levels that a magistrates court could award for pollution from £2000 to £20 000 in 1991, magistrates seemed swayed more by arguments of mitigating circumstances focused on the farmer's (lack of) culpability than by the consequences of the pollution itself. Thus, in court, Pollution Inspectors faced a system of judgement in which the moral discourse focused on the farmer and not the pollution, and fines were usually low. The average fine for farm pollution in the South West region in 1990 was £425, with 63 per cent of those prosecuted receiving fines of less than £500. Such low levels of fines were seen by Pollution Inspectors to discourage farmers from taking care to prevent pollution and to perpetuate a discourse of pollution as merely overstepping the rules. The following year, many fewer prosecutions were taken (26 compared to 40 in 1990), but the average fine level rose to £665, reflecting the view expressed by one senior pollution control officer that it was not worthwhile to pursue cases that would probably attract fines of below £500.

Pollution Inspectors were also frustrated that more of the pollution incidents they presented for prosecution were not pursued in court, ascribing this in the main to the requirement for clear-cut "watertight" cases. Despite their heightened sensitivity to pollution and the instigation of legal training for new staff, it was clear that Pollution Inspectors found problems in switching to a more legally oriented style of regulation and in getting the details right. For those who had worked for the former regional water authority, legal action and specifically formal sampling were neither part of their routine work culture nor of their training. Some cases had failed because of insufficient or incorrectly (according to judicial rules) gathered evidence by Pollution Inspectors. This, in turn, reinforced a cautiousness on the part of senior staff in deciding to initiate prosecutions. They only wished to proceed with cases they were completely sure of winning. It was preferable to issue a warning letter or a caution if there was any chance of losing a case. Indeed, the National Audit Office found that the NRA's prosecution success rate was some 97 per cent in 1993 (National Audit Office 1995: 2).

Above all, Pollution Inspectors held strong reservations about the efficacy of legal action. On the one hand, low fines seemed no deterrent; on the other, they were concerned lest a more coercive stance on their part might damage their relations with the farming community. Most still felt that the prosecution of farmers had less value in preventing pollution than did friendly persuasion. Although they recognized that there was some leeway to tighten up on the prosecution of uncooperative "problem" farmers (and here the implicit or explicit *threat* of prosecution was seen as a useful device), they displayed ambivalence about the value of prosecuting "persuadable" farmers or pushing any to the extent that they went out of business. Although they recognized serious pollution as an environmental crime that should be punished, the orientation of Pollution Inspectors' preventive strategies around the assessment of farmer types was disturbed when it was a "persuadable" rather than a "problem" farmer whom they were faced with prosecuting. The crime may be worthy of punishment but not necessarily the perpetrator. Without the cooperation of the large majority of farmers, the Pollution Inspectors' task would have been impossible, and in order not to alienate them, Inspectors were encouraged to act in a way that the farming community regarded as reasonable. Such an enrolment was eased by the agricultural connections or training of several Inspectors, their interaction with ADAS officials and, most important, the everyday contact that Pollution Inspectors had with farmers and their pollution problems.

The farmers' networks

The farming community is often characterized as being relatively closed because of the extensive overlap in social and kinship networks that tend to set farmers apart (Newby et al. 1978). Their closest friends are often other farmers, as are their closest relatives. Social change in the countryside, while eroding that closedness, is also reinforcing for many farmers their sense of being different and separate. For some, this means an absorption into non-farming networks; for others, a closing in amid a growing feeling of beleaguerment; and overall, a fragmentation of the farming community. Pollution regulation, as we have seen, is tied up with social change. The responses of farmers to pollution problems and pressures do differ and can be understood in terms of the different networks in which they are involved in a changing rural society.

These differences are highlighted if we compare the farmers at either end of the spectrum of opinions on farm pollution and their social networks. The middle ground of opinion (just under two-thirds of our sample of farmers) held that pollution was a significant problem for farming, which, although not necessarily the farmers' fault, should be tackled to avoid trouble and to protect the image of agriculture. However, there was also a group of farmers (just one-fifth of the sample) who were unequivocal in viewing farm pollution as reprehensible (we have labelled these the "radicals"); and another group (one-sixth of the sample) who questioned the "fuss" being made about it (we have labelled them the "sceptics"). The radicals were, on average, somewhat younger farmers who occupied larger farms, compared to the sceptics. The radicals tended to have more complex land occupancy or tenurial relations: whereas only a third of them ran family owner-occupied farms, all of the sceptics did. The radicals were generally less committed to family continuity on the farm and were much more likely to have embraced diversification as a farm survival strategy, which in several cases had included enterprises associated with countryside recreation or tourism. A majority of the radicals also had rivers or streams running close to their farms, and some of them used or let the fishing rights.

By drawing comparisons between the two extreme groups, we can begin to see how a farmer's networks might shape the way in which pollution problems were perceived (Ward & Lowe 1994, Ward et al. 1995a). The sceptics who felt that farm pollution was not a serious problem could be said to show characteristics of a "traditional" approach rooted in the ethos of family farming. They saw themselves as a special group in society, set apart from everyone else. Family farming dominated their lives and they argued that "farming is all we know". Thus, they tended to be more locked into a productivist agricultural way of thinking, and were much less likely to have diversified their farm businesses. They tended to have reactive as opposed to proactive farm management strategies. Changes on the farm during the 1980s were more likely to have been agricultural and made in response to changing economic or family circumstances. These farmers were more likely to feel embattled because of farming's poor economic fortunes, and environmental and pollution regulations were seen as yet another attack on farmers. However, notions of family continuity remained strong among this group and were linked to dynastic notions of land ownership and farm improvement. The sceptics tended to belong to local farming families with close relatives farming nearby, which may have reinforced their inward-looking and agro-centric view of the world.

On the other hand, the radicals who saw farm pollution as reprehensible could be said to show characteristics of a more "modern" approach. First, that they tended to be younger farmers is important because, not only did they tend to be trained to a higher level, but their experience of training was more recent. Also, they tended to have a broader outward-looking view, associated with their more extensive social and economic links beyond farming, including links with the non-farming public through on-farm diversification and an inclination to socialize much more outside the local farming community. These farmers, in enjoying wider links beyond farming, were in turn reflecting the views about farming's environmental problems held in wider society. They did not tend to see themselves as a special group set apart. With regard to their own farms, these farmers tended to be less risk averse and more flexible. They viewed themselves more as rural entrepreneurs or "businessmen". Changes in the farm business were often proactive, and they were more likely to have established non-agricultural enterprises on their farms. At the same time, they were losing the dynastic sense of family continuity and, on several farms, succession to the next generation had definitely been ruled out. Some of the farmers in this group had young children who were learning about environmental issues and problems at school, thus providing another route by which new sets of environmental values "flowed" into the household.

The ways in which farmers in different social networks had come to view pollution differently are illuminated in the following case study examples of a sceptic, a radical and a farmer representative of the middle ground of farming opinion (we have labelled this middle group the "ambivalent farmers").

Case study 1: farmer Hill (a "sceptic")

Mr Hill is in his late-50s and he runs a 60 ha dairy farm with his wife and three sons. The farmyard and buildings are tidy and well maintained, and lie at the end of a long track surrounded by rolling hills. Mr Hill comes from a large local farming family and has several other close relatives involved in farming in the area. Explaining that the farmwork takes up all his time, Mr Hill remarks that he has "no social life". The family does not travel much: only the odd agricultural show every few years takes him out of Devon. The family are very much part of the farming community and tend to have little to do with "the outside world".

The farm business includes beef and sheep enterprises and has been built up in recent years in preparation for two of the sons, who are in their teens and are keen to enter the family business. Expansion in the 1980s continued a trend evident since the 1940s. The milking herd quadrupled in size from 20 cows between 1950 and 1980, and grew by a further ten during the 1980s, bringing the total by 1991 to 90 cows. Additional milk quota was bought in the late 1980s to facilitate this continued expansion. The beef enterprise has also been built up and 20 ha of additional land has been acquired. Mr Hill, indeed, equates "improvement" with "expansion". Much of his farming life has been spent under a system designed to encourage farmers to expand the scale of their production, and it is probably his enthusiasm for those past clear goals that makes adjusting to new conditions all the more difficult. Nevertheless, being a progressive farmer is an important element of Mr Hill's approach: "if you are a good farmer, you are going to have to be up to date". Although the farm's profitability has fallen slightly in the past two years, Mr Hill has no specific plans to improve its economic performance, seeing the next few years as a period of consolidation. When asked if he plans to diversify, he replies "Good God, no! It's too good a farm to do that", the implication being that diversification is an option forced upon failing farmers on "poor" farms.

For Mr Hill, farming is under attack in several ways. He sees the efforts of farmers, trying to get on with what they know best, as being constantly encroached upon by external meddling: "the attitude of the media and government policy makes you feel cynical. And you get the city coming out to the countryside and telling us what to do, like on Exmoor with the stag-hunting. City people criticize [us] but the farmers know best". When it comes to farming's environmental problems, Mr Hill feels that the "over-use of fertilizers" could affect the environment, but he rejects criticism of pesticides with the argument that: "You've got to use sprays to keep yields up". Increasing environmental concerns have not affected him. When asked if he has experienced any pressure from local people to alter his farming practices, he replies "No, I'm out of the way", and expresses sympathy for other local farmers who have received complaints for walking stock on local roads. "The townies in the village haven't got a clue", he remarks.

Mr Hill's dairy cows are housed in cubicles and the amount of slurry produced has increased as the herd has grown. He had an above-ground tank installed in the 1970s and has the capacity to store slurry for three months without spreading. In 1990 he added a system of pumps to handle the dirty water. As long as the drainage channels are kept clear from blockages and the pumps are in order, the system works well. Mr Hill is also aware of the risks of pollution from spreading slurry on land. As he explains, "You have to make sure that the ground will take it and it won't run off". However, there could be problems if he were required to upgrade his waste facilities, for example to enlarge its storage capacity. He has recently borrowed money to buy additional land, and the thought of having to find additional cash to spend on waste control makes him shudder. "The government would have to pay for it", he declares.

Public concern over farm pollution, Mr Hill feels, has got out of perspective. He characterizes "pollution" as "interfering with the natural environment in an unjust way", adding illuminatingly that "nature can cope with so much muck, but not too much". He feels that pollution from farms is much less significant than from industry, but that despite this, farmers are being more strictly regulated. "We are not the main culprits. I suppose with the misuse of chemicals and fertilizers . . . yes . . . but with slurry in water, it's not a chemical. It is not a dangerous substance. It's got to go somewhere. It's got to go into Mother Earth" he explains. In any case, the extent of farm pollution ought to be diminishing. "We should see more life in the rivers in the future than over the last few years. Intensification is slowing down." In dealing with agricultural pollution, he admits "you've got to have guidelines" but regulation shouldn't get "too official". He goes on to explain "the law's the law, but there's always an odd occasion when there's an accident. You get unforeseeable bursts. Even with [above ground] stores, they have a lifespan. You get these natural accidents". Under such circumstances, prosecution would be unfair.

To summarize, Mr Hill's approach to farming has been shaped by his experience of more than 40 years of intensification and expansion. The farm has done reasonably well and, fortunately, as part of its general modernization, the waste facilities have been adequately upgraded, and so the risks of serious pollution are fairly low. This is all the more fortunate because Mr Hill has some trouble seeing farm pollution as a significant environmental threat. His social networks rarely extend beyond the farming community and this reinforces his view of pollution being a problem brought upon farmers from elsewhere. For him, pollution control is part of the regulatory onslaught that makes life ever more difficult for farmers like himself, who just want to get on with the job of producing their quota of milk as efficiently as possible. It is those farmers with similar perceptions of farm pollution, but with inadequate waste storage and disposal systems, that can pose the greatest problems for the regulators.

Case study 2: farmer Cross (a "radical")

Mr Cross, who is in his early 30s, joined his family's farm business in 1981 and is now effectively the principal manager of the 100 ha dairy, beef and arable farm, which is mostly rented. The dairy herd has fallen in size by about 20 per cent over this period. The family is not totally dependent on agriculture for its income, and there is a possibility in the future of expanding the modest holiday cottage business, but this would depend on planning permission being granted. The holiday cottages bring tourists and visitors onto the farm, and Mr Cross's networks extend well beyond just the farming community. Discussing farming issues with friends and holiday visitors has given him a strong sense of the ways that farming can be misunderstood by the non-farming public. He explained that the farming community needs to be "a little bit more careful about the view we portray to the general public. Farmers have been poor purveyors of what we do. With the scandals there's been an element of worry but they haven't been *explained* well enough".

Mr Cross sees himself as "a perfectionist rather than an expansionist", and explains "we cut the herd partly as a result of quotas, and partly because we can produce the same quantity of milk from a smaller number of cows . . . I get more pleasure from doing it right rather than getting bigger". Indeed, he rates "enjoying the farming way of life", "looking after the countryside" and "keeping the land in good heart", as being more important priorities than "maximizing profitability" or "being a progressive, up-to-date farmer".

This does not mean that he is backward-looking or resistant to change, simply that he strives for an integrated view of the development of the farm that encompasses the wellbeing of his land and animals, as well as his own personal satisfaction in the ways things are done. That tends to make him sceptical of external advice and critical of fellow farmers who blindly follow it. He sums up his own approach to farming as "being very critical of what we do". In comparing his own farm work with surrounding ones, he assesses, in particular, "the quality of people's successes and mistakes". He feels that his own empirical expertise is more important than the external advice available to which he seldom turns. In the 1980s, for example, he switched completely from hay to silage, primarily for the labour-saving advantages. Now between 30 and 40 per cent is maize silage, which suits the rotation, the feeding system and the local climate better, and he is proud of his achievements in increasing the proportion of home-produced feed. His own farm trials led him to ignore the advice of the chemical companies on fertilizer application, and he now applies no manufactured nitrogen to the maize, although four bags were recommended. He sees high input use as a sign of bad farming, and emphasizes that "it's not margin per hectare that matters. Any berk can make a profit with 500 cows on a small acreage".

To him, the art of good farming is mixed farming. He takes satisfaction from the fact that it requires a wider range of skills and specialist knowledge, and he feels strongly that "mixed farming helps the landscape".

Environmental problems arise because "farming has become over-intensive in certain areas". His own environmental concerns have affected his farming practices. He cites two examples. First, the family considered switching to a slurry-based system with cubicles, but the potential for slurry problems persuaded them to stay with loose housing. Mr Cross felt that in doing so, the comfort and happiness of his cows were better ensured and that using more solid manure would benefit the land. Secondly, by increasing the acreage of maize for silage, Mr Cross feels that "we now don't have to push the grass so hard", meaning he need not apply such high levels of inputs and there is less pressure to maximize the number of cuts for silage.

Mr Cross feels strongly that "pollution controls have to be taken on board, particularly on livestock farms", and sees this as one of the main issues confronting dairy farming. He feels that statutory controls on waste management are the right way forward. He is sceptical about voluntary changes in farming practice based on government advice, partly because such advice is often too abstract or misguided. In an area of under 75 mm rainfall per year and with a non-slurry-based housing system, he recognizes "it's a relatively easy job for me". The solid muck gets mixed with straw, stored on concrete and then ploughed in. He also gets about one tonne, or one rota-spreader load of slurry, per day from scrapings from the collecting and silage yards, which is always spread on land that will be ploughed for maize. However, there are still risks. Yard washings are not separated from rainwater and slurry has to be spread every day through the winter ("something that will have to be changed in the future").

He is careful not to make any silage when it rains or if rain is forecast and he aims for high dry-matter silage. He feels that "MAFF advice that direct-cut silage was better than dried silage was one of the worst pieces of advice they ever gave". His own approach means no significant effluent is produced. As with other technical shortcuts, he is dismissive of silage additives: "if weather conditions are such that you need to use an additive, you shouldn't be making silage".

The two case studies illustrate how the values of the farmers in the two "extreme" categories differ in terms of pollution issues, and farming problems more generally. But as farmers' values change over time, how is the make-up of these groups likely to change? We would expect that in the future those values that tend to be held by the "radicals" will become more widespread, whereas the number of farmers sharing the views of the "sceptics" will shrink. Social change in the countryside is likely to confront more and more farmers with the attitudes of the non-farming public. Also, greater awareness of pollution and other environmental issues is likely to continue to filter through to farms via a variety of channels ranging from changing syllabuses in agricultural colleges to television programmes on the environment. Of crucial importance will be the response of the broad middle group of farmers. An insight into some of the pressures they face and how they are coping is provided by our third case study.

Case study 3: farmer Jones (an "ambivalent" farmer)

Mr Jones is in his mid-40s and he farms an 80 ha dairy farm with his wife. He told us how he is becoming worried about changes in his small hamlet. Until the late 1980s, it was a farming settlement, but in recent years the two other farmhouses have been sold off and a neighbour's barn has been converted into residential units. With his farm being the one remaining and with eight private dwellings close by, Mr Jones fears that there will be opposition to ordinary farming practices from the new-comers to the hamlet. He wondered, "will they object to the noise and the smell?" How will they react to the noise of the milking machinery or the cattle moving around the yard in the early morning? He is fearful of objections, and complains about "more town people in the countryside putting pressures on farmers . . . They don't like to see mud on the road and the like".

Mr Jones's response to successive external events – milk quotas, BSE, the storms of 1987, farm pollution controls and social change in the village – has been to try to absorb them but not to be deflected. The prime consideration has been to build up and maintain the dairy business as a viable unit in terms of its demands on family labour and the income that it provides. As Mr and Mrs Jones approach middle age, their ability to cope with these demands has become an important consideration. During the past ten years, they have built up the dairy herd, and have acquired additional quota and land, with the aim of achieving and then maintaining an 80 plus milking herd. However, external events seem to be conspiring against this strategy. Mrs Jones, in particular, has begun to question the likelihood and even desirability of their son taking over the business in the future, and to explore alternative sources of income.

In coping with external shocks, Mr Jones has developed a risk-averse approach to farming. Before the imposition of milk quotas in 1984, he attempted to build up the herd, when "everyone was encouraged to expand".

Although since then he has increased the dairy herd to 90 cows by buying and leasing quota, he does not consider this to have been an expansion. Rather, it is a response to having BSE in the herd and, he feels, is a case of "running to stand still". The disease tends to appear when a cow is at its fullest milking potential, and so can have a disproportionate impact on the productivity of a herd.

The latest external shock is the new pollution regulations. Their effluent management system was upgraded in 1990. In addition to the slurry pit, with its ten-week storage capacity, a dirty water system now collects rainwater and runoff from the yard and spreads the effluent on the land. Mr Jones stresses that there is a running cost associated with the new system. "Every time it rains we've got a two-horsepower motor running." More work is also involved: the dirty-water sprinklers have to be regularly moved by hand. Conversely, Mr Jones perceives few, if any, benefits: there has been only a minimal reduction in nitrogen use on the land, and the soil has become saturated.

The £21 000 for the dirty-water system was a major capital outlay for the Jones family, even if half was covered by a grant. The prospect of such an outlay, which, as they saw it, would not contribute to their overall strategy (of making the holding more viable as a working family-run unit), catalyzed a reassessment of that strategy. First, it led the family to look at radical change to preserve the strategy and then, when this did not seem possible, to begin to question the strategy itself.

They had other capital priorities and it did not seem right to sink so much into a farm where much of the animal housing was in need of modernization. Following the lead of a neighbouring farmer, therefore, they first commissioned a feasibility study on the possibility of converting the barns, selling the farmhouse and moving to a new one with modern buildings elsewhere on the holding. However, the costings showed a £100 000 shortfall and they were not prepared to go that much into debt. The neighbour, though, has gone ahead and, having obtained planning permission, is now selling out completely with the probability of another new non-farming neighbour for Mr and Mrs Jones. "Farmers are being forced out of business by pollution regulations", he complains. "They know they will have to spend £20 000 and many see it as their last chance to sell up with milk quota."

The concatenation of external events has thus begun to shake their resolve. Mr Jones looks in vain to government to provide some stability but instead sees it as a destabilizing force. Indeed, he feels that the major difficulties facing farmers today are the "uncertainty in policy; of government and the EC not planning far enough ahead. With a dairy farm you want to know what you're doing in three years time." Pollution regulations not only seem to be the last straw but have brought home to them that they are caught up in a larger power struggle, over which they have no control. Ultimately, it may engulf them, making it difficult or impossible to continue farming in the way they have been.

Although the national and European policy contexts seem so uncertain and subject to remote decision-making, the most immediate threat to the continuity of their farming operations is on their very doorstep. Here, the microgeography of the farm and its neighbours is of crucial significance, with the prospect of the farm being the only one left in the small hamlet. "It's the new people in the countryside who will be imposing the regulations", Mr Jones comments resignedly. With a sense of the external pressures closing in on them, he and his wife have begun to question their farming strategy, with Mrs Jones taking the lead. Part of the house has been converted for holiday letting; 1990 was the first year. Mr Jones suggests that the income from it has been modest so far. Even so, he has extended a wet clay pit and stocked it with tench and bream to provide fishing with the holiday accommodation, in order to extend the letting season. Bringing tourists and visitors onto the farm to fish was also having the effect of extending the Jones's networks beyond the farming community, and anglers in particular bring with them quite different experiences of farm pollution from those of the farming community. Although the other half of the longhouse is derelict, there is the possibility of repairing it for holiday accommodation or long-term letting. But Mr Jones warns "it's not the golden egg". It would cost a lot to repair.

When asked what his aims for the farm business are, Mr Jones refers to the uncertainty over milk quotas. Are they here to stay? Will they remain transferable? Will they be cut further? A 2 per cent cut in quota would oblige them to lease 50 000 litres of quota, as against 30 000 currently, just "to stay still". Alternatively, "we could cut the herd by about ten" and switch resources to repair the house. This he identifies as the key decision they will have to make in the next two years. He has also heard rumours of the possibility of even tougher pollution controls and that may make it prudent to curb the size of the herd. Although they seem beset by insecurity and uncertainty, there is a tangible sense that they are approaching a crossroads in their farming endeavours.

Conclusions: farm pollution and patterns of enrolment

Although pollution is measurable in terms of its impacts on aquatic life, we have seen in this chapter that what counts as pollution is socially defined. Therefore, just what pollution *is*, and how much is acceptable, become key questions. Our study found different definitions of pollution and different understandings of the seriousness of the problem. There was a reasonable consistency of views about what constituted pollution between regulators and advisers in the field, and a variety of views among farmers. However, there were considerable disparities between the groups over the tolerance of pollution. ADAS advisers were much more willing than NRA Pollution Inspectors to tolerate higher levels of pollution risks, because of the widespread lack of financial resources on farms to invest in pollution control, whereas farmers were tolerant of "accidental" pollution.

Environmental activists were very intolerant of farm waste pollution, which they saw as the outcome of agriculture's recklessness. Increasingly, this view was reflected in the media reporting of farm pollution incidents and helped shape the formal regulatory regime that was introduced in 1989. However, although water pollution was classified as a criminal offence, the view of farm pollution as "environmental crime" was not dominant among the field-level actors. Some of the NRA's informants took this view and it was becoming the orthodoxy among NRA regulators, but quite different perspectives prevailed among advisers, farmers and magistrates.

Thus, the fundamental struggle centred on whose representation of the pollution problem should prevail. The basic conflict was between the definition of farm pollution as a problem *for* farming, which implied agriculturally led solutions (such as upgrading storage facilities according to the needs of the current production system), and farm pollution as a problem *of* farming, which implied environmentally led solutions (such as the setting of water quality standards, with changes being made to production systems as appropriate). Of course, underlying the distinction, between pollution as a technical problem (a form of "rule-breaking") and pollution as something discreditable that attracts blame (a type of "environmental crime"), was whether or not any moral opprobrium attached to it.

Among the dairy farmers we interviewed, there was the full spectrum of views – from perceptions of agricultural waste as "natural", to rule-breaking, to environmental crime. This represented a spectrum from those who emphasized the productive context in which pollution arose to those who emphasized its environmental consequences (in other words, from an inward-looking to an outward-looking perspective). For many of the farmers who viewed pollution as a form of rule-breaking, contextual questions to do with intentionality were clearly of significance in judging any particular incidents (was the pollution a deliberate, wanton, or an accidental act?). Such judgements focused on the responsibility of the actor rather than the consequence of the action. This was a discourse that farmers' leaders had helped to structure. It was one that incidentally, happily saw them translated as instruments of government policy – any responsibility for pollution when farmers were conforming to policy or following official advice was thus displaced.

It seems that ADAS regarded pollution in a matter-of-fact manner. It was a problem that needed to be dealt with, but it carried no moral overtones. This could be because as engineers they generally adopted a pragmatic problem-solving approach to the succession of issues that confronted farmers, and could also be because past ADAS advice was so implicated in the causes of pollution. In many instances, the judgement ADAS staff were making was essentially a balancing of the risks of pollution and detection against the financial costs (of prosecution or remedial action) to the farmer.

The NRA was confronted by a much more complex set of judgements. Traditionally, the Pollution Inspectors of the former water authorities had treated pollution as a technical issue. However, the establishment in 1989 of the NRA as a national regulatory body, cast by itself and the environmental lobby in the role of environmental watchdog, injected a new and morally charged discourse into the proceedings. The Pollution Inspectors had thus to operate day-to-day in a farming world (and arguably in an institutional tradition) in which pollution was predominantly considered a technical issue. However, the NRA increasingly adopted a public stance that water pollution was unacceptable and reprehensible.

More generally there were now two conflicting forces at work. A general increase in interest in environmental issues and the rising numbers of articulate middle-class people who had moved to the countryside accounted for the growing public intolerance of farm pollution. Images of farming as the "natural" way of life were easily shattered by the noise of automated milking parlours, the smell and ubiquitous nature of slurry, and the overall factory-like appearance and practices of modern industrial farms. Public opinion had come to have a growing influence on the water authorities, particularly with pollution increasingly defined by members of the public reporting pollution incidents they had encountered. Furthermore, the creation of the NRA had not only given institutional expression to this popular concern but had also significantly altered the balance of expert opinions. In particular, the domination of the water authorities by chemists and engineers had been superseded by a rising concern with biology and the conservation of complete water systems. This holistic view of the management of water environments enlarged the definition of pollution and underpinned a lowering of tolerance thresholds.

At the other extreme, the views of agriculturalists on the definition of pollution and its seriousness, which in the past had dominated the debate[2] still continued to have great influence. The notion of "good agricultural practice", which until the 1990s had defined "good environmental practice", was still prevalent. Despite increasing environmental concerns and surpluses of many agricultural products, faith in high-tech intensive farming had not been undermined. From the approaches and attitudes adopted, it would appear that the productivist era had not ended for most agriculturalists and farmers.

The distinction between pollution as a technical problem and pollution as environmental crime encompassed two quite separate patterns of enrolment among the various actors involved. Technical problems were appropriately dealt with through adjustments to farming practices, effected by advice, information, the formalization of standards and incentives; this was the realm of the Farm Waste Regulations, capital grants and ADAS advice. Criminal acts, in contrast, elicited remonstration, public condemnation and prosecution; this was the realm of pollution incidents, environmental campaigning and the courts. Arguably, in the implementation of the Regulations, ADAS and the farming community were seeking to enrol the NRA into a technical definition of the problem, with the interpretation and implementation of the Regulations and the Farm and Conservation Grant Scheme central to the process of enrolment. On the other hand, environmental groups and a wider public were seeking to enrol the NRA into treating pollution as an environmental crime, with the emphasis on the recording, investigation, reporting and prosecution of pollution incidents. Central to the enrolment process here was the ability of a member of the public to report a pollution incident, which then had to be investigated. The availability of published information on incidents then gave the opportunity for environmental groups both to express alarm about the scale of the problem and to condemn what could be presented as leniency.

It is notable how unequal these enrolment processes were. It was farmers who had to be persuaded or dragooned into acting differently. Their limited scope and inclination for action was a major determinant of the pace of change. In addition, the Pollution Inspectors were in day-to-day contact with the farming community. In contrast, environmental groups were only fitfully involved in the issue. The agricultural community was thus much better placed to impose its definitions of the problem and to proffer solutions on a day-to-day basis.

As we saw earlier, the availability of a means to quantify farm pollution had helped turn it into a public problem. The efforts of water regulatory staff to quantify farm pollution had not only made it into a public problem but also had encouraged its popular definition. Crucially, farm pollution incidents depended upon members of the public reporting them and, from the mid-1980s, the regional water authorities and subsequently the NRA had actively encouraged the public to do just that. The NRA had continued to rely on the level of incident-reporting as an indicator of the overall problem, as well as in its day-to-day work in tracking down sources of pollution. The NRA, with its telephone hotline for the reporting of pollution incidents, clearly regarded this public responsiveness as both a boon to its operational duties and as a means of maintaining popular support for its work. What it did not welcome, though, was the way in which the details of incident statistics and prosecution levels allowed environmental activists both to interfere with its operational activities and to challenge publicly its commitment to a strict approach towards farm pollution. These concerns prompted the NRA to revise the way it reported pollution incidents and prosecution totals. The overall effect was to de-emphasize farm pollution and to obscure its prosecution practices.

Although the NRA thus sought to diminish the influence of environmental campaigners on its work, it also sought to extend its own influence over farm management and agricultural policy. The instrument for achieving this aim was the farm waste management plan. The recognition by MAFF of such plans as a potential tool in the minimization of farm pollution signalled, at the policy level, the acceptance of the NRA as a member of the farm waste policy community and, at the operational level, the establishment of a modus vivendi with ADAS. By 1994 the pilot scheme in which ADAS helped farmers prepare farm waste management plans had been extended to cover 16 river catchments. MAFF reported the results of an ADAS evaluation of the scheme, which found that, of the farmers volunteering to take part, 50 per cent were able to draw up their own plans with little assistance from ADAS and that over 80 per cent of the participants had incorporated the plan into their waste management practices in some way (MAFF 1994a). Although these voluntaristic pilot schemes fell well short of the NRA's ideal of compulsory waste management plans for all farms subject to its sanction and subordinate to catchment management planning, what they did do was to begin to integrate waste management into farm management and to implicate ADAS advice in the responsible disposal, as well as the storage, of farm waste. In this way, the onus of responsibility for the preventive management of farm waste was pinned on farmers and their advisers and was hopefully made a feature of routine farming practice rather than being seen as something which could be solved simply by buying new storage facilities or the latest spreading equipment. Crucially, though, the approach adopted remained firmly within the voluntaristic ethos which had come to characterize British agri-environment policy.

With the overall problem of farm waste pollution seemingly being contained, the NRA was content not to press the issue further. The DOE set the seal on this settlement between the NRA, ADAS and MAFF when it responded belatedly in 1995 to a major Royal Commission report on freshwater quality that had been published in 1992. The Royal Commission had proposed a much more protective approach to water quality, one that would involve less reliance on the capacity of the water environment to deal with wastes. It therefore endorsed the NRA's earlier proposal for a comprehensive system of farm waste management plans, drawn up within a catchment planning framework. In response, though, the DOE confirmed what was now the official line, that encouragement of voluntary farm waste management plans was the most suitable and "practical" means of addressing pollution risks from farm waste (DOE 1995: 28). However, with waste management planning becoming an accepted feature of agricultural advice and farm management, a bridgehead had been established for a dirigiste approach to be resurrected, should matters deteriorate and the policy climate change.

However, MAFF was strongly committed to the voluntaristic approach and was concerned to show the effectiveness of its grant aid scheme and advisory efforts. Increasingly it had expressed confidence that the farm waste problem was being effectively addressed. By the end of 1993 about 9000 farmers in England had received grant aid for pollution control measures. According to the Minister of Agriculture, this had "led to a steep drop in the number of major agricultural pollution incidents" (MAFF

1993). The government felt able then to cut the level of grant aid provided to farmers from 50 per cent to 25 per cent, and a year later it was abolished. The press release that accompanied the announcement of the ending of the grant proclaimed that "in England the number of major pollution incidents caused by agriculture had fallen from 99 in 1991 to 67 in 1992 and to 63 in 1993" (MAFF 1994b).

Had the problem been solved then? Politically it had. The announcement of the abolition of the grant scheme caused little reaction from farming or environmental organizations. But had it been solved practically? This question is far more difficult. The fundamental pressures on farmers and their farming practices had not changed significantly. There had certainly been some improvement in the storage and handling of farm waste, but only about a third of dairy farmers had made grant-aided improvements to their facilities. However, no-one could now have been unaware of the farm pollution problem and the possibility of prosecution for causing it. Unfortunately, the change in the basis of the statistics does not allow clear inferences to be drawn on what the consequences were for pollution incidents. From 1991 onwards, only the categories of major and substantiated incidents were publicly reported, but neither of these categories had been in use previously, with the result that meaningful comparisons over time could not be made. The following years' figures do indicate a modest decline in *major* farm pollution incidents. At the same time, however, the number of *substantiated* incidents caused by farming was rising. Rather incredibly, the proportion of major to substantiated incidents fell from 7.6 per cent in 1990 to 1.1 per cent in 1994. At such a tiny and diminishing percentage, to place great store by precise absolute numbers of major incidents and the apparent trends they reveal would demand an unwarranted faith in the complete consistency and objectivity of scores of Pollution Inspectors across the country. Unfortunately, the Pollution Inspectors were under complex and changing institutional pressures. Increasingly, it was in their own personal or organizational interest to be very cautious in judging a pollution incident to be a major one. If they did so, they personally lost the discretion that they normally enjoyed in responding to an incident or in dealing with a farmer. If an incident was categorized as major, and then subsequently a prosecution was not pursued or did not succeed, then this could be judged a personal and an organizational failure. Overall, there was an organizational imperative to match the number of major incidents recorded with the number of prosecutions pursued and to do so in a context from 1992 onwards in which the NRA informally de-emphasized prosecution as a means of tackling farm pollution. All one can say with any certainty is that the problem of gross pollution from farm wastes was statistically buried.

Starved of much of the information that had made campaigning possible and which previously had traced a telling upward curve, environmental activists lost interest. In any case, other issues had overtaken farm pollution. The interest in it had been more about its *symbolic* significance than its *practical* consequence; the symbolism had related to concerns about agriculture's environmental responsibilities and the development of official regulatory strategies at a time of water privatization. By the early 1990s, other issues more powerfully captured these concerns and attracted the attention of environmental groups. The focus of their campaigning was increasingly international, especially at the European level. For agriculture, this meant such matters as the Habitats Directive, the Nitrate Directive and CAP reform. Environmental groups were attracted to the European Union (EU), not only because this was where the political action was in the development of environmental policy and legislation, but also because the style of EU regulation, with its explicit formal standards and reporting requirements specified in law, gave them a firmer purchase on environmental problems and their regulation than did the informal practices that British procedures nurtured (Lowe and S. Ward 1998).

As the focus of attention moved on, where did this leave the farmers? They continued trying to earn their living through their dairy farms. Most had come through this episode chastened and with a greater understanding and sense of responsibility for the problems caused by the management and disposal of their farm waste. Undoubtedly, the increased attention given to prosecutions in the late 1980s and early 1990s had been a crucial factor. But its significance to the farmers was misunderstood, and this pointed to a wider gulf of misunderstanding that served to isolate the farming community. In our survey we came across no farmers who consciously or deliberately took the sort of calculated risk over the possibility of prosecution that ADAS advisers presumed they did, or over the scale of the fines following prosecution that Pollution Inspectors assumed they did. The reason why none of the farmers thought in this way was because all of them regarded the prospect of prosecution not as an affront to their wallets but as an affront to their good name and character. In general, they saw the threat of prosecution and the publicity such cases attracted in the local press as personally demeaning and as a slight on the industry. It contributed further to their sense of alienation and beleaguerment. However, for the farmers too, the salient issues moved on, as new controversies emerged around the transportation of live calves, and fears of BSE. For many, though, these served to reinforce a sense that farmers were the victims of the prejudices and whims of a demanding but uncaring urban society. Farming organizations pointed to the grim statistics, indicating higher suicide levels among farmers, to dramatize the pressures and isolation that some farmers face. Pollution Inspectors were not unmoved by such worries and anxieties. Indeed, we were told, without any apparent sense of irony, by one of the NRA officials who had helped set up the pollution hotline, that he had subsequently been asked to be involved in the setting up of a Samaritan advice line for despairing farmers.

Notes

1. The unattributed quotes given in this section are taken from our interview with Andrew Hickson, the Devon FOE campaigner (13.8.91).
2. So much so that for many years there was no debate.

CHAPTER EIGHT

Conclusions: constructing moral orders

Introduction

In Chapters 3 and 4 we saw how farm pollution in Britain shifted from being a "non-issue" during the 1970s, contained within the realms of the agricultural and water policy communities and treated as little more than a technical side effect of efficient production, to a situation where it became seen as a pressing environmental problem, a key contributor to declining river quality and an issue demanding a much stronger regulatory approach on the part of government. In Chapters 5 and 6 we moved away from Whitehall and Westminster and the "high politics" of farm pollution, where policies are formalized, to the worlds of NRA Pollution Inspectors and dairy farmers deep in rural Devon. In this final chapter, we return to a discussion of the overall analytical approach to our study, which we outlined in Chapter 1. In presenting our conclusions, we attempt to situate our study within wider theoretical questions in social science, not only about agriculture, pollution and regulation, but also about what we have termed the *moralization* of the environment.

This chapter turns first to a discussion of the different views of pollution and the two moral discourses that surround them. It then goes on to examine the implications of our study on our understanding of what German sociologist, Ulrich Beck, describes as the risk society and, in particular, the role of regulatory science in bringing environmental problems to light. Thirdly, it returns to a discussion of actor-network theory to illustrate how various roles and representations became ascribed and fixed in the networks through which farm waste pollution became recognized as a public problem and hence an object of regulation. The chapter ends with a section that focuses on the key field-level relationship between the farmer and the Pollution Inspector, as a site where the two competing moral discourses had to be reconciled in everyday life.

Nature, rurality and morality

Chapter 1 introduced a distinction between two views of pollution: one seeing it as a technical problem (a form of "rule breaking") and the other seeing it as something discreditable that attracts blame (a type of "environmental crime"). A major achievement of the contemporary environmental movement, we would argue, has been to establish an abstracted conception of pollution as wrongdoing. It has thereby elevated pollution and industrial risks to the status of a crime, and forged a new environmental morality.

Our study of the *moralization* of the environment has taken the example of farm pollution as its focus. However, this has proved not to be a simple tale of public-spirited campaigners stigmatizing a hapless group of producers for the environmental damage they cause. What we saw instead was two moral discourses pitched against each other: the new environmental morality encountering an older rural morality rooted in the virtuous industry of the farming community.

When defending themselves against charges of damaging the environment, we found that farmers were able to draw on a rich repertoire of justification, which essentially naturalized their actions, by portraying themselves as subject to elemental forces that they could only partially control and which therefore absolved them of culpability. In the past, somewhat similar arguments have been used to characterize groups such as children, women and native people as creatures of nature, not entirely responsible for their actions and therefore not capable of being full moral subjects. Although many farmers do think of themselves as victims (or as being victimized), they also see themselves as proud, independent countrymen who are inferior to none. To reinforce this identity, they can draw on traditional notions of masculine authority, particularly those of the toiling provider and the guardian of the countryside. Such rhetorics are often expressed by the self-same farmers who also claim diminished responsibility for slurry leaks and spills, occasionally in the self-same breath.

Traditionally, these obverse images – of farmers as both objects and agents in the interface between society and nature – have had wide currency.

The sons of the soil were also the husbanders of the land. This complex dual imagery served to represent farmers as *natural subjects*, playing a vital role in both nature and society, and with such responsibilities – to the land, to the countryside, to socially vital needs – that transcended mere material interest or social convention.

At the root of the farmers' responses was a set of distinctions that equated agriculture with nature and distinguished it from industry, in denying or minimizing the pollution risk from farming. In doing so they were drawing on long-established categories. In Chapter 3 we saw how these categories had structured official action over many years. When, in the 1970s, legislation was passed to tighten up pollution control, agriculture was formally exempted. Later that decade, the Royal Commission on Environmental Pollution felt impelled to investigate agriculture, but even then emphasized agriculture's fundamental distinctiveness deriving from its embeddedness in nature. Indeed, the Royal Commission proposed that those husbandry activities which it recognized as significant sources of pollution – namely intensive livestock rearing – be redefined as not agricultural but industrial. Agricultural pollution was thus not only an aberrant act but also an anomalous category.

Where did these established categories come from? The distinction between what is natural and what is unnatural is a fundamental one in the construction of different social orders. It was Mary Douglas, in her study *Purity and danger* (1966), who first drew attention to the importance of the maintenance of boundaries between categories such as these. In seeking to understand why something perfectly acceptable to one society or social group is seen as dangerous to another, she pointed to the ways we make sense of the complexities of the world by categorizing phenomena and drawing morally charged boundaries between them. Things that transgress, or destabilize, conventional boundaries are often seen as threatening, or "polluting". In Douglas's celebrated phrase, pollution is "matter out of place".

We can see how industrial society was constructed on a sharp, if ultimately elusive, distinction between nature and society (Latour 1993). Whereas science and technology sought to subjugate nature, the industrial city separated people physically from the natural world. Rural areas, stripped of most of their population and their preindustrial economic activities, became specialized spaces for producing food for the urban population. In this way the rural became equated with the agricultural. But the rural also acquired strong associations during the period of industrialization: it was seen as where many city dwellers had come from, and for many it also represented the last vestiges of their contacts with nature. Rural and agrarian values thus figured strongly in cultural reactions to the disruptions and depredations of urban industrialization and became a central feature in the definition and assertion of national identities against what was seen as the rootless and homogenizing cosmopolitanism of the industrial age (Lowe & Bodiguel 1990, Lowe et al. 1995). Rurality was thus portrayed as "natural" and it became a source of moral affirmation and condemnation (Bell 1994). In contrast to the supposed innocence of rural life, the city was perceived as corrupting, not only of traditional morality and social hierarchies, but also of nature, through industrial pollution (Mumford 1961). We thus see a correspondence between the categories of society and nature, urban and rural, and industry and agriculture.

Foucault (1965, 1970) suggested that industrial society sought systematically to manage a form of "social order" through the imposition and maintenance of certain predominant categories such as these. Other important ones were those of male and female, home and work, and capital and labour. Crucially, these categories reinforced one another, and together formed a means of systematic "ordering" (Law 1994). By providing a place for everything, they helped to keep everything in its place, and thus to establish the natural order of things. Through such interlocking dichotomies and their morally prescribed boundaries of what was natural and what was unnatural, identity and social behaviour became fixed in a social structure.

Agriculture was an important receptacle of the natural morality. The discourse that projected natural values onto farming and rurality originated with poets, philosophers, naturalists, artists and the like, not with country people who were often portrayed as rural innocents (i.e. natural subjects). Real agriculture was always more complex, including both technologically progressive "improving" farmers and backward or traditional "dog and stick" farmers. Nevertheless, the discourse of natural morality was available for use as a resource by agriculturalists in promoting and defending agriculture. For example, the discourse has been drawn upon to emphasize the farmers' responsibility for the countryside, with the implication that environmental concerns around agriculture are best left to self-regulation and voluntary schemes. Equally, the equation of agriculture with the rural and the natural on the one hand, and of pollution with the industrial and the urban on the other hand, for a long time precluded the very category of agricultural pollution from recognition. Conversely, when agricultural pollution did come to be recognized, it was a pointer not only to the changing nature of agriculture but also to wider shifts occurring in social meaning and sources of authority, including the erosion of some of the basic dichotomies of industrial society. What has happened in recent years is that, while nature continues to be a source of moral values and an inspiration to efforts to sustain or recreate a new moral order, the actors that now successfully mobilize this notion are *environmental* groups and activists rather than agriculturalists. In consequence, contemporary conflicts between farming and the environment have a wider significance than the specific issues of concern, being also struggles between competing claims to authority. Indeed, agricultural pollution emerged as an issue in the 1980s out of the highly charged debates over who should be trusted with the vital responsibility of safeguarding the purity of water.

The moralization of risk and regulatory science

Throughout this period, what became apparent was a growing public sensitivity towards risk, not least the risks involved in any contamination of water sources. In recent years, the most influential sociological characterization of technological and environmental risks has been that outlined by Ulrich Beck (1992a,b). According to Beck, the distinctive nature of contemporary risk is that it arises from the techno-economic decisions and considerations of utility made by commercial and state organizations. It is this that makes industrial risk a political issue. He goes on to argue that a series of contemporary technological developments pose acute difficulties for the political management of risk because of the scale, ubiquity, incalculability or unaccountability of the dangers they pose. This is exacerbated by the "monopoly of scientists and engineers in the diagnosis of hazards" (Beck 1992b: 107) including often the self-same people who are responsible for creating the hazard in the first place. Beck's proposed solution is a challenge to technocracy – a call for the establishment of democratic control over "the hazard producing and administering industries" (ibid.: 115). The agents for this "ecological extension of democracy" (ibid.: 119) must be oppositional citizens groups, since business, science and the state are too implicated in the technological project. In this unequal struggle, the main weapon citizens' groups have at their disposal is the contradictions arising from a potentially "self-annihilating progress" (ibid.: 113).

Our own study confirms much of Beck's analysis but challenges it in crucial respects. In particular, our study demonstrates how regulatory officials and citizens groups can combine to establish pollution as a major public issue open to popular definition and with science playing a supporting role. In seeking to build a constituency for public regulatory functions at the time of water privatization, pollution regulators were able to draw on the discourse of environmental morality propagated by the environmental movement. In doing so, they needed to publicize and help politicize the existence of significant but overlooked problems of pollution. Farm waste pollution exactly fitted the bill. However, in order to define this matter as a public problem, there was a need for data. Hence farm pollution incident statistics were first collated, and these seemed to indicate a large and growing problem (Lowe et al. 1996).

What happened in the politicization of farm pollution, though, differs from standard accounts of scientification (i.e. the process whereby a public problem is redefined as a technical or scientific problem). Undoubtedly, many individuals and organizations in the past had suffered from farm pollution (anglers, riparian owners, other farmers, fish farmers, water suppliers, water consumers, naturalists and so on). With the exception of the "bad smell", however, the issue had not been a public problem but a technical problem (i.e. one subject to engineering or agronomic definitions and solutions). Previously, the role of science – at least that conducted by agricultural researchers on odour measurement and control – had been to belittle the public aspect of the problem.

Now, in contrast, the availability of a means to quantify pollution helped turn it into a public problem. Rather than expert closure, the very opposite occurred. The attempt to quantify farm pollution, orchestrated by water regulatory staff, not only made it into a public problem, but also encouraged its *popular* definition. Crucially, farm pollution incident statistics depended upon members of the public reporting incidents and, from the mid-1980s, the regional water authorities and subsequently the NRA actively encouraged the public to do just that. The NRA continued to rely on incident reporting as a gauge of the overall problem, as well as in its day-to-day work in tracking down sources of pollution. The NRA clearly regarded this public responsiveness as both a boon to its operational duties and as a means of maintaining popular support for its work. There are parallels in other fields, such as the relationship between voluntary wildlife groups and conservation scientists and officials – a network which, through its ability to assemble diffuse evidence about the status of wildlife, has been able to uncover various insidious environmental threats including, for example, the link between the use of certain pesticides and declines in bird populations (Moore 1987). Such examples suggest that, through strategic links between regulatory scientists and environmental groups, particular sociocultural constructions of risk can be established to accommodate popular experience.

Networks, enrolment and identity creation

These examples also illustrate how accounts of pollution problems generally involve a host of social actors, natural entities and technical artefacts, and so pose a challenge for conventional sociological study. The story of farm waste pollution, for example, requires that we consider the roles and functions of farmers, environmentalists, Pollution Inspectors, magistrates, cows, fish, snails, slurry, rain, pits, tanks, pumps and various pieces of paper (forms, guidelines, regulations and codes of practice). The recognition of the problem and its solution necessitate the orchestration of these diverse and diffuse elements and actors. That is, they are the outcome of actors constructing networks. In the words of Callon et al. (1985: 10), the role of sociological analysis within an actor network approach is to study the creation of "categories and linkages, and examine the way in which some are successfully imposed while others are not". The key to the approach is the way in which actors create networks of relationships that assign roles and identities to others. Through the successful creation of networks, some actors are thereby able to assert particular meanings and representations over others.

On the one hand, the broad recognition of farm waste pollution involved the creation of a network drawing together environmental groups, regulatory officials, policy-makers, parliamentarians, farmers and concerned members of the public. That network was able to establish the notion of farm wastes as a major source of water pollution in the late 1980s. In doing so, it established new identities for the various actors involved – for example, casting members of the public as pollution watchdogs ("the eyes and ears" of the regulators in the words of the NRA), farmers as potential polluters, and the NRA as "Europe's strongest environmental protection agency". On the other hand, the regulation of farm pollution involved another network which, although intersecting with the first, was separate from it. This second network encompassed the field-level actors: the farmers, ADAS officials and Pollution Inspectors. It presented the solution for farm pollution as being a matter of improved and expanded storage of farm wastes. It cast farmers as waste managers and the field-level officials as advisers in the storage and disposal of farm wastes.

These were not the only efforts at enrolment that were taking place. For example, environmental activists sought to enrol Pollution Inspectors, regional NRA managers, farmers, the local media and magistrates to adopt a view of pollution incidents as criminal acts demanding summary prosecution. At the same time, farmers' leaders with policy-makers, and individual farmers with the officials they encountered, sought to recruit support for their own self-image as responsible producers and countrymen attempting to cope with a regulatory onslaught.

How then do some networks succeed over others? If actors act in accordance with the roles assigned to them, then this strengthens a network. For example, the NRA's acceptance of the mantle of environmental watchdog was in keeping with the role assigned to it by the environmental lobby. Likewise, many farmers tacitly accepted that they were potential polluters. On the other hand, despite their detailed involvement with the farming community over farm waste management issues, NRA Pollution Inspectors acquiesced in the role assigned to them by ADAS officials of being specialists in river pollution but inexpert in agricultural matters. Similarly, it was the view among farmers and agricultural advisers, encouraged by MAFF's Farm and Conservation Grant Scheme, that the "pollution problem" could be "solved" through investment in technical equipment to store farm effluents and dispose of them to land; a grant awarded and equipment installed signified a problem solved. NRA officials struggled against this technical-fix representation. In emphasizing the need for farmers to carry out routine monitoring and maintenance of waste facilities and to be vigilant against pollution, they sought to enlarge the accepted notion of responsible farming practice. When such ascribed roles were accepted, and actors acted accordingly, networks were strengthened and order emerged.

Networks eventually become stabilized through the circulation of "things" between actors. As Callon remarks "actors define one another in interaction – in the intermediaries that they put into circulation" (Callon 1991: 135). Law suggests that some materials and artifacts appear more *durable* than others and so maintain their relational patterns for longer. "When we start to *perform* relations – and in particular when we *embody* them in inanimate materials such as texts and buildings – they may last longer" (Law 1992: 387). Hence a good ordering strategy, in the building of stable networks, is to embody a set of relations in durable materials. Durability may be a function of how intermediaries are constructed or what they are constructed from.

Legislation is one way of constructing durable intermediaries. The Farm Waste Regulations, for example, established a statutory role for Pollution Inspectors in overseeing farmers' waste facilities. Likewise, when the *Code of good agricultural practice* changed from being a defence of farmers to a defence of the environment, its legal basis was reconstructed. Conversely, the NFU and MAFF resisted any suggestion that farm waste management plans should be made legally compulsory. Another way of stabilizing networks through intermediaries is to fix relationships technologically (i.e. to embed them in equipment or set them in concrete). The most durable practical legacy of the episode we have studied was undoubtedly the many tanks, pits and pumps that farmers were induced and subsidized to acquire. These technological artefacts cast farmers as waste managers. They illustrate Latour's claim that "technology is society made durable" (Latour 1991).

A variety of intermediaries, including pollution incident statistics, telephone hotlines, regulations, codes of practice and waste management plans, were all important in efforts to mobilize diffuse networks, in order to stigmatize or regulate farm pollution. However, their meanings were not predetermined, but were derived from, and reinforced by, particular patterns of enrolment. Separate networks thus construed them differently, and the ambiguities in the key intermediaries in our study underlined the fact that one network had not prevailed over all others.

A pollution hotline, for example, could be a means of mobilizing popular support for the pollution regulators; however, environmental activists might seek to use it to gain leverage over the regulators' response to pollution incidents. A Pollution Inspector's signature on a farmer's grant application form for a pollution control scheme also signified quite different things to different actors. To the NRA and to ADAS, it meant that the scheme would be *technically* adequate to address the problem; but ADAS staff saw this as a formality if they had designed the scheme; and the NRA staff stressed the additional requirement of continual and careful maintenance and management. To many farmers, though, the NRA's approval signified that the NRA was implicated in the decision about how best to address pollution risks through technology: any future pollution problems would, therefore, be "its fault too". Another example is the *Code of good agricultural practice*, which was cast as a farmers' defence under a former pattern of enrolment; but came to be seen as an environmental defence when a new pattern was established. Likewise, waste management plans were put forward by the NRA as part of its development of a strategic and preventive approach to catchment management planning and regulation, one that would cast the farmers partly in the role of "waste treatment engineers"; but waste management plans were taken up by ADAS and MAFF as a means of extending the voluntary approach to agri-environmental policy. These ambiguities represent attempts at counter-enrolment and serve to indicate that the predominant pattern of enrolment is rarely total or complete.

The key actors in these competing networks understood that they were involved in a struggle to establish or resist different meanings. Even when closure had been effected, there were still lone voices who sought to challenge a now predominant meaning. Some farmers, for example, did not think of farm wastes as potent pollutants and, in refusing to identify themselves as potential polluters, were inclined to ignore all those intermediaries that would tend to cast them in such a role, including the Farm Waste Regulations, the *Code of good agricultural practice* and the Farm and Conservation Grant Scheme. Likewise, the NFU would not accept pollution incident statistics as a measure, however crude, of the incidence of farm pollution. Instead, and as if taking a leaf out of actor-network theory, they objected to these statistics on the grounds that they were a social construction (because of their dependence on public reporting)!

Agriculture's moral economy

In responding to charges of pollution, farmers were not abject; nor was their sense of being responsible producers and countrymen a sham. In understanding their responses, we have drawn on the notion of a *moral economy* (Scott 1976), by which we mean a community's working definitions of what is fair and unfair.

At the time of our field study in Devon, pollution was certainly among the most pressing issues that dairy farmers faced nationally. The problem was seen by farmers as yet another issue in a succession of crises – including the imposition of milk quotas, the storms of 1987, BSE and so on – seemingly visited on them by a capricious world. Most accepted the desirability of avoiding pollution of rivers by farm wastes and felt they had done what they could to minimize pollution risks. Mainly, they had done so within what might be termed their "productivist rationale" which, at the same time, acted as a major constraint in really tackling the problem. Understanding productivism is essential to understanding how farmers view the environment and how the farm pollution problem often poses such conceptual difficulties for them.

Society needs food and so depends upon the success of farmers in carrying out their work. This fact provided the obvious and undeniable foundation for portraying agriculture as a vital and socially worthwhile activity. The experience of urban industrialization and the consequent spatial and occupational separation of the bulk of the population from its own basic provisioning served to heighten this sense of dependency, particularly in the century of total war and the threat this posed to the food security of urban populations. But the rationale of the sector was not simply to provide, but became, above all, to expand. The philosophy of productivism construed production as good, but also more production as always better.

Productivism resonated with a work ethic which saw sloth rather than greed as sinful, linked virtue and industriousness, and saw greater wealth as the just reward for hard work. Thus, *productive* farmers were seen as *better* farmers, with greater yields displaying the virtuousness of hard work. This work ethic became especially associated with agrarian culture, particularly since urban culture had come to project virtues such as refinement, taste and artistic achievement, which bespoke leisure time, and vices, such as unemployment, welfare dependency and hedonism, which bespoke idleness and dissolution.

The quest to "make two blades of grass grow where only one once grew" became agriculture's central maxim in the technological revolution of the post-war decades. It infused agricultural policy, but became even more firmly ingrained at the farm level, where technological changes brought profound increases in output and productivity. The effect was to eliminate the old dichotomy between progressive and traditional farming and replace it with a single dominant model of "technocentric productivism". Paul Thompson describes such a philosophy as "the headlong and unreflective application of industrial technology for increasing production" (1995: 70, see also Ward 1993, 1995). As such, it could almost be described as an anti-environmental philosophy, save for one important fact. Productivism in agriculture was usually combined with and qualified by the duty of stewardship – the responsibility to care for nature. This notion was often expressed by farmers in terms of keeping the land "in good heart", including a sense that sound farming involved passing on the farm to the next generation in "a better condition". Farmers also readily acknowledged a wider role that they played within the countryside, and often talked of their contribution to the making of the landscape. They used farming's past contribution as a rhetorical defence against contemporary accusations of polluting nature.

However, according to Thompson, traditional agrarian stewardship "is conceived as a duty ethically subservient to production; hence when stewardship would entail constraints on production, duties to nature seldom prevail over the productivist ethic" (1995: 72). In effect, modern technology drove a wedge between the farmer's interest in production and traditional agrarian stewardship, although within the moral economy of the farmer the discourse of stewardship is still drawn upon. However, the crucial distinction between productivism and stewardship as a set of values is that the former is embodied in the techno-economic system within which farmers are embedded, whereas the latter is individualized and seen as a matter of personal responsibility.

Thus, few of the farmers we surveyed were critical of technology. They tended to see technological change as inevitable and ineluctable, closely associated with notions of "agricultural development" and "progress". The vast majority of farmers did not see pollution as an indictment of the farming systems or the technologies they used. Although technology was thus seen to be an impersonal but benign force, pollution problems were seen to arise from changing public attitudes. Increased public sensitivities to environmental issues, coupled with a climate of opinion less sympathetic towards farming, was leading to growing intolerance of the problems farmers faced. On this issue, therefore, farmers' personal sense of worth (in the stewardship ethos) faced social rejection, and inevitably this reinforced their feeling of isolation and alienation.

The farmer and the field-level bureaucrat

The farmers were not the only ones in our study who had to deal with contradictory demands. This was also true of the Pollution Inspectors. They operated at the intersection of the two main networks we have identified: one constructed on the basis of environmental morality; the other on agriculture's moral economy. Other studies of the working lives of local officials have characterized their positions as ones of structural ambiguity or contradiction in having to mediate between grassroots realities and centrally imposed rules. Here, two such literatures have proved helpful in interpreting our findings.

The first literature originates in studies of the implementation of public policy (particularly urban and social policy) and concerns the work (in exercising discretion, interpreting policy and in dealing with the complexity of the social world) of the so-called "street-level bureaucrat" (Lipsky 1980). Given the different context of our work, we have adapted this label to that of "field-level bureaucrat". The second literature originates in development studies and rural sociology, where "actor-oriented" perspectives have been developed to examine "encounters at the interface", typically between rural producers and representatives of states, development agencies or technology interests (see, for example, Long 1989).

The street-level bureaucrat literature tells us that "the decisions of street-level bureaucrats, the routines they establish, and the devices they invent to cope with uncertainties and work pressures, effectively become the public policies they carry out" (Lipsky 1980: xii). Such a perspective suggests caution over the feasibility of top-down policy change amid the inertia of established ways of working and field-level relationships. It is through implementation "on the street" (or in our case, in the field) that *de facto* regulatory policy is created (see also Ham & Hill 1993). Of crucial importance is the development by street-level bureaucrats of practices that enable them to cope with the pressures they face. Lipsky explains, "they believe themselves to be doing the best they can under adverse circumstances and they develop techniques to salvage service and decision-making values within the limits imposed upon them by the structure of their work. They develop conceptions of their work and of their clients that narrow the gap between their personal and work limitations and the service ideal" (Lipsky 1980: xii). The field-level bureaucrats we studied conformed to this model. The literature suggests their coping strategies may include: choosing easy rather than difficult cases; routinization of procedures and working methods; standardized classification of the regulated world and of client groups; and adopting a cynical attitude to ambitious goals and their replacement with more personal goals.

The literature also identifies a range of regulatory policy styles, with legislative approaches at one extreme and *laissez faire* (with its associated risks of policy "capture") at the other. Between these two extremes lies "flexible enforcement" (see Bardach & Kagan 1982) where discretion is exercised over centrally devised rules, but where the identification of "good apples" and "bad apples" among target groups comes to be of crucial importance. We found this to be so in our study of Pollution Inspectors and farmers.

The basic twofold division which Pollution Inspectors make of farmers – that of "persuadable" and "problem" types – fits in well with findings from other studies of regulatory agencies. These have recognized that, whereas officials regard most offenders as potentially compliant, there is usually a small minority who cannot be persuaded to comply with the law by appeal to a sense of social responsibility (Hawkins 1984). However, in none of these other studies does there appear to be such suspicion of the regulated group to the extent that an overriding model of the regulated as potential law-breakers emerges.

Pollution Inspectors have an *absolute* notion of farm pollution. To pollute a river or stream is a wrongful act, because of its effects. Pollution Inspectors have an expert understanding of the sensitivity of the water environment and the functioning of aquatic ecosystems, and they are aware of the potency of farm effluents. They therefore fully appreciate the damage that farm pollution can do. Their task is to "make rivers clean" and they are reluctant to regard *any* amount of pollution as acceptable.

In contrast, farmers have a *relative* notion of pollution informed less by an understanding of its consequences and more by an understanding of the context in which it arises. Dairy farmers have to deal with considerable volumes of farm effluents every day. In heavy rain or when storage facilities fail, effluent escape is unavoidable, and may be mitigated by the circumstances – for example, the rapid dilution and dispersal of pollutants in a river in spate. Of course, a large spillage of effluent is regrettable but "a little won't hurt"; after all, farm wastes are "natural" substances. For farmers, *the morality of farm pollution concerns the morality of the deed* – whether the pollution was deliberate or accidental. Their sense of personal worth, responsibility and circumstances all come into play around the morality of farming and the morality of pollution.

In recent years, farmers have been required to alter their farming practices, sometimes in major ways, because of pollution regulations, but these changes go "against the grain" of the prevailing production logic. They have been reluctant to spend what they regard as "dead" money on installing pollution equipment, and have resented the additional regulatory powers being exercised over them. Many farmers in our survey saw these regulations as indicative of a wider "anti-farmer" sentiment in society. In part, this explains their rather begrudging response. As one farmer succinctly explained to us:

> Farmers are more aware of the problem and are making plans to put it right. [Pollution regulations] haven't made people change their set ups too much. They just try to deal with it in their own way. For example, people haven't switched to straw bedding or cut their stock. They try and go on as they were and do what they have to, to stop pollution. (farmer interview)

That is to say, when farmers did take steps to improve pollution control on their farms, they were "taking it on board" but not "altering course". The fact that intensive slurry-based dairy farming continued as the model production system was, however, increasingly at odds with the growing public perception, propagated by environmental pressure groups, of farm pollution as an indictment of modern farming practices. Most of the changes made on farms were undertaken within the constraints of *food production* processes, rather than as part of the farmers' role as *environmental managers*. Our survey revealed a few farmers making changes in their fodder cropping patterns and feeding practices, for example, in an effort to make slurry more manageable and so reduce the risk of pollution. However, such steps remained the exception rather than the rule, and tended to be limited to those farmers with both the resources and an understanding of the scale of the challenge that pollution might pose for the farming industry. For the majority of farmers, modifications to their pollution-control facilities and procedures were largely reactive, made in response to regulatory pressures from the field-level bureaucrats, and were mainly "add on" changes that did not alter their production systems.

Although the Pollution Inspectors were armed with new powers, the farmers were not powerless in the face of a seemingly much stricter regime. After all, it was they who had to be persuaded or dragooned into acting differently. Their limited scope or inclination for action was a major determinant of the pace of change. In his study of forms of resistance among peasants in rural Malaysia, Scott examined the ordinary, everyday "weapons" of relatively powerless groups in the face of state bureaucracy or capitalist exploitation (Scott 1985). Scott's subjects, as subordinate classes, were interested less in changing the larger structures of the state and the law than in what Hobsbawm (1973) termed "working the system . . . to their own minimum disadvantage". In such an endeavour, the weapons at the peasants' disposal included foot-dragging, false compliance, feigned ignorance and so on. In similar terms, what "weapons" do British dairy farmers use when confronted by new regulatory controls?

We have identified four types of strategy from our fieldwork. The first involved "keeping your head down", hoping no problems would arise that would draw the regulators' attention to your farm, and hoping the issue of pollution would just "go away". The second strategy was to pursue delaying tactics in order to avoid spending on pollution control facilities or to deflect prosecution. This might involve stringing out negotiations with Pollution Inspectors, losing forms, pleading lack of time or resources, arguing that any required action was inopportune, or endless procrastination. The third strategy was to seek to coopt the Pollution Inspectors. This might involve the farmer in trying to explain the specific difficulties posed by the farm circumstances, such as the poor financial conditions or the level of indebtedness, in order to "educate" the Pollution Inspectors about the particular constraints faced. Farmers also treated Pollution Inspectors as advisers, thereby seeking to implicate them in the decisions and actions taken. A fourth "weapon" of resistance was to attempt to "blackmail" the Pollution Inspectors. These attempts ranged from threats that the farm might go out of business if regulations were enforced, to threats of noncooperation, and even to threats of violence. In our interviews with farmers and our participant observation studies "shadowing" regulatory officials in their work, we learned of examples of each of these strategies.

In response to the variety of farmers, and because of the farmers' ability to exercise sanction over regulatory power, whether through inaction or malicious action, the Pollution Inspectors had to develop coping strategies. Often, these strategies involved the Pollution Inspectors developing working classifications of farmers as "good" and "bad". Such categorizations were relative, distinguishing between conscientious or "persuadable" farmers and "rogue" or recalcitrant ones. Crucially, the Pollution Inspectors' categories resonated with those of the farming community in being based on the moral worth of the farmer rather than the consequences of the pollution.

Usually, Pollution Inspectors' weapons of environmental morality were thus only deployed when they corresponded to the farmers' own moral values. The only farmers automatically prosecuted in the courts for causing pollution were those responsible for major, serious pollution incidents (that is, in obeisance to environmental morality). Beyond that, the only farmers to be prosecuted were those viewed essentially as "rogue" farmers, not only by the Pollution Inspectors but also by the farming community. It was as if prosecution had either to have the tacit acceptance of the farming community or it had to be obviously seen by the Pollution Inspector to reinforce the categories held by the farmers themselves – in other words between what was seen as responsible and what was seen as utterly irresponsible behaviour. Overall, under 10 per cent of farm pollution incidents were usually prosecuted each year. Thus, although Pollution Inspectors were armed with the outlook and powers of the new environmental morality, they found in their efforts to improve river quality in practice that they were dependent upon the cooperation of the farming community. To secure that cooperation they had to work with the grain of what that community considered fair and unfair, right and wrong – in other words, the farming community's moral economy. Punishing farmers who deliberately emptied slurry into rivers would be acceptable, but it would be less acceptable to prosecute where overflows or equipment failure had occurred in heavy rain, or where farmers were doing their best to improve waste management facilities, or where enforcement action might put them out of business.

Afterword

The politicization of farm pollution led to the imposition of various controls on agriculture to restrain the excesses of agricultural productivism. What these environmental regulations meant in practice was that farmers were increasingly confronted by regulatory officials armed not only with new powers but also with a new moral authority. The exchanges that took place between them were not limited to what constituted sound agricultural practice, but touched on nature, morality and the law. Much of what was of interest to us not only in the regulation of polluting agricultural practices, but also in terms of the making and undoing of moral discourses, occurred in these encounters. What was revealed was a traditional order (of agricultural productivism) under challenge and a new order (around environmentally responsible agriculture) being discursively constructed and resisted.

As we put the finishing touches to the manuscript for this book, we cannot help but reflect upon how our account of the moralizing of the environment might be relevant beyond the confines of our particular empirical focus of farm pollution. For the past nine months of 1996, British agriculture has been in turmoil over BSE, following the acknowledgement by the Secretary of State for Health in March 1996 of the possible link between eating contaminated beef and the identification of a new strain of the human brain disease, Creutzfeldt Jacob Disease. For the Devon dairy farmers we studied, the scale of this crisis will have swamped the farm pollution crisis of a few years earlier. In a perceptive piece in *The Independent*, journalist Andrew Marr has pointed to the ways that the degree of public outrage over BSE reflect a clash of moral values (Marr 1996). Mad cow disease is not just a question of public health, nor simply a question of regulatory politics: "It is a question of town and country. It has provoked the biggest crisis for generations in relations between farming . . . and the cities where most of us live." Marr's story is a familiar one, and similar to our own. Farmers were once "respected national heroes whose sweat and knowledge helped the country survive war and eat ever better in peacetime". Yet recently we have seen a dramatic change of attitude. "A rampant urban moralism has increasingly painted farmers as big business villains – cruel, greedy, insensitive, polluting". Campaigners protest about live animal exports, suspicious consumers view meat and fruit with fear and trepidation. "Urban Britain stares bleakly at rural Britain, and finds it wanting. . . . From the other end of the telescope, the minority still working the land stares back at the cities and suburbs and sees a haze of hypocrisy and ignorance".

We have sought in our study to reveal the roots of these competing moralities by examining how farm pollution shifted from being a technical side-effect of efficient food production to an environmental crime. We have done this by tracing the views and actions of those actors most closely involved, following them as they constructed their networks, precariously building their worlds and struggling to convince others of what is natural and what is unnatural, what is right and what is wrong.

Bibliography

Advisory Council for Agriculture and Horticulture 1975. *Inquiry into pollution from farm waste, part III – report on pollution from farm wastes* (December). London: Ministry of Agriculture, Fisheries and Food.

Agricultural Research Council 1976. *Studies of farm livestock wastes.* London: Agricultural Research Council.

Bailey, R. & J. Minhinick 1989. The agricultural requirement for water, with particular reference to irrigation. In *Agriculture and the environment: technical papers from the annual symposium,* Institution of Water and Environmental Management, 8.1–8.16. London: Institution of Water and Environmental Management.

Bardach, E. & R. Kagan 1982. *Going by the book: the problem of regulatory unreasonableness.* Philadelphia: Temple University Press.

Beck, L. 1989. A review of farm waste pollution. *Journal of the Institution of Water and Environmental Management* **3**, 467–77.

Beck, U. 1992a. *Risk society: towards a new modernity.* London: Sage.

Beck, U. 1992b. From industrial society to the risk society: Questions of survival, social structure and ecological enlightenment. *Theory, Culture and Society* **9**, 97–123.

Bell, M. M. 1994. *Childerley: nature and morality in a country village.* Chicago: University of Chicago Press.

Bolton, N. & B. Chalkley 1990. The rural population turnaround: a case study of North Devon. *Journal of Rural Studies* **6**, 29–43.

Brassley, P. 1996. Silage in Britain, 1880–1990: the delayed adoption of an innovation. *Agricultural History Review* **44**, 68–87.

Brownlie, T. & N. Taylor 1982. *Report on survey of fodder silos.* Edinburgh: Department of Agriculture and Fisheries for Scotland.

Callon, M. 1986a. Some elements of a sociology of translation: domestication of the scallops and fishermen of St Brieuc Bay. In *Power, action and belief: a new sociology of knowledge,* J. Law (ed.), 196–233. London: Routledge & Kegan Paul.

Callon, M. 1986b. The sociology of an actor-network: the case of the electric vehicle. In *Mapping the dynamics of science and technology*, M. Callon, J. Law, A. Rip (eds), 19–34. London: Macmillan.

Callon, M. 1991. Techno-economic networks and irreversibility. In *The sociology of monsters*, J. Law (ed.), 132–61. London: Routledge.

Callon, M., J. Law, A. Rip 1985. How to study the force of science. In *Mapping the dynamics of science and technology*, M. Callon, J. Law, A. Rip (eds), 3–17. London: Macmillan.

Centre for Agri-food Business Studies 1991. *A survey of dairy farmers' attitudes towards MMB reform*. Royal Agricultural College, Cirencester.

Champion, A. & A. Townsend 1990. *Contemporary Britain: a geographical perspective*. London: Edward Arnold.

Clark, J. & P. Lowe 1992. Cleaning up agriculture: environment, technology and social science. *Sociologia Ruralis* **32**, 11–29.

Clark, J., P. Lowe, S. Seymour, N. Ward 1994. *Sustainable agriculture and pollution regulation in the UK*. Working Paper Series 13, Centre for Rural Economy, University of Newcastle upon Tyne.

Commoner, B. 1966. *Science and survival*. New York: Viking.

Confederation of British Industry 1987. Memorandum of evidence to the Committee. See House of Commons Environment Committee (1991) 63–7.

Conway, G. & J. Pretty 1991. *Unwelcome harvest: agriculture and pollution*. London: Earthscan.

Council for the Protection of Rural England 1987. Memorandum of evidence to the Committee. See House of Commons Environment Committee (1987) 360–66.

Cox, G., P. Lowe, M. Winter 1988. Private rights and public responsibilities: the prospects for agricultural and environmental controls *Journal of Rural Studies* **4**, 323–37.

— 1990. *The voluntary principle in conservation*. Chichester: Packard.

Department of the Environment 1976. *Pollution control in Great Britain: how it works* [Pollution Paper 9]. London: HMSO.

— 1983. *Agriculture and pollution: the Government's response to the seventh report of the Royal Commission on Environmental Pollution* [Pollution Paper 21]. London: HMSO.

— 1987. Memorandum of evidence to the Committee. See House of Commons Environment Committee (1987) 1–14.

— 1988. *The Government's response to the third report of the Environment Committee (Session 1986–87) on pollution of rivers and estuaries* [HC Paper No. 543]. London: HMSO.

— 1995. *Freshwater quality: Government response to the sixteenth report of the Royal Commission on Environmental Pollution*. London: Department of the Environment.

Department of the Environment/Welsh Office 1986a. *River quality in England and Wales 1985: a report of the 1985 survey*. London: HMSO.

Bibliography

Department of the Environment/Welsh Office 1986b. *The water environment: the next steps*. London: Department of the Environment.

— 1992. *Planning policy guidance: the countryside and the rural economy* [Planning Policy Guidance 7]. London: HMSO.

Douglas, M. 1966. *Purity and danger: an analysis of the concepts of pollution and taboo*. London: ARK Paperbacks.

ENDS 1985. MAFF persuaded to tighten agricultural pollution code. *ENDS Report* **120**, 21–2. London: Environmental Data Services.

ENDS 1990. River authority wants farm waste plans to reduce water pollution. *ENDS Report* **183**, 6–7. London: Environmental Data Services.

Federation of United Kingdom Milk Marketing Boards 1991. *United Kingdom dairy facts and figures – 1991*. Thames Ditton: Federation of United Kingdom Milk Marketing Boards.

Foucault, M. 1965. *Madness and civilization*. New York: Pantheon.

— 1970. *The order of things*. New York: Vintage.

Friends of the Earth 1987. Memorandum of evidence to the Committee. See House of Commons Environment Committee (1987) 160–96.

— 1991. *The water campaigners guide to farm pollution*. London: Friends of the Earth.

Furness, G., D. Colman, S. Webb, K. Hendry 1991. *The status of waste handling facilities on livestock farms in Great Britain, 1990/91*. Report to the Ministry of Agriculture, Department of Agricultural Economics, University of Manchester.

Gowan, D. 1972. *Slurry and farm waste disposal*. Ipswich: Farming Press.

Grove-White, R. 1993. Environmentalism: a new moral discourse for technological society? In *Environmentalism: the view from anthropology*, K. Milton (ed.), 18–30. London: Routledge.

Grundey, K. 1980. *Tackling farm waste*. Ipswich: Farming Press.

Halliday, J. 1987. *The effect of milk quotas on milk producing farms: a study of registered milk producers in the Honiton and Torrington areas of Devon*. Exeter: Devon County Council/University of Exeter.

— 1988. Dairy farmers take stock: a study of milk producers' reactions to quota in Devon. *Journal of Rural Studies* **4**, 193–202.

Ham, C. & M. Hill 1993. *The policy process in the modern capitalist state*, 2nd edn. Hemel Hempstead: Harvester Wheatsheaf.

Haslam, S. 1990. *River pollution: an ecological perspective*. London: Pinter (Belhaven).

Hawkins, K. 1984. *Environment and enforcement: regulation and the social definition of pollution*. Oxford: Oxford University Press.

— 1989. Rule and discretion in comparative perspective: the case for social regulation. *Ohio State Law Journal* **50**, 663–79.

Hawkins, E., J. Bryden, N. Gilliatt, N. MacKinnon 1993. Engagement in agriculture 1987–1991: a West European perspective. *Journal of Rural Studies* **9**, 277–90.

Hays, S. 1984. The British conservation scene: a view from the United States. *Ecos* **5**(3), 20–27.

Hellawell, J. 1986. *Biological indicators of freshwater quality and environmental management*. London: Elsevier.

Hill, M., S. Aaronovitch, D. Baldock 1989. Non-decision-making in pollution control in Britain: nitrate pollution, the EEC Drinking Water Directive and agriculture. *Policy and Politics* **17**, 227–40.

Hobsbawm, E. 1973. Peasants and politics. *Journal of Peasant Studies* **1**, 3–22.

Holmes, N. 1990. British river plants. *British Wildlife* **1**, 130–43.

House of Commons Committee of Public Accounts 1991. *Advisory services to agriculture* [HC Paper 465, Session 1990–91]. London: HMSO.

House of Commons Environment Committee 1987. *Pollution of rivers and estuaries* [HC Paper 183–I, Third Report, Session 1986–7]. London: HMSO.

Howarth, W. 1992. Agricultural pollution and the aquatic environment. In *Agricultural conservation and land use: law and policy for rural areas*, W. Howarth & C. Rogers (eds), 51–72. Cardiff: University of Wales Press.

Institution of Water and Environmental Management 1987. Memorandum of evidence to the Committee. See House of Commons Environment Committee (1987) 243–8.

Kinnersley, D. 1988. *Troubled water: rivers, politics and pollution*. London: Hilary Shipman.

— 1994. *Coming clean: the politics of water and the environment*. London: Penguin.

Kneale, J., P. Lowe, T. Marsden 1992. *The conversion of agricultural buildings: an analysis of variable pressures and regulations towards the post-productivist countryside*. Countryside Change Working Paper 29, Department of Agricultural Economics and Food Marketing, University of Newcastle upon Tyne.

Knowland, T. 1993. *Changing the guard? institutional change in water pollution control*. PhD thesis, School of Environmental Science, University of East Anglia.

Latour, B. 1986. The powers of association. In *Power, action and belief: a new sociology of knowledge*, J. Law (ed.), 196–233. London: Routledge & Kegan Paul.

— 1987. *Science in action: how to follow scientists and engineers through society*. Milton Keynes, England: Open University Press.

— 1993. *We have never been modern*. Hemel Hempstead: Harvester Wheatsheaf.

Law, J. 1992. Notes on the theory of the actor-network: ordering, strategy and heterogeneity. *Systems Practice* **5**, 379–93.

— 1994. *Organizing modernity*. Oxford: Basil Blackwell.

Lipsky, M. 1980. *Street-level bureaucracy*. New York: Russell Sage.

Long, N. (ed.) 1989. *Encounters at the interface: a perspective on social discontinuities in rural development*. Department of Sociology, Wageningen Agricultural University, The Netherlands.

Lowe, P. & M. Bodiguel (eds) 1990. *Rural studies in Britain and France*. London: Pinter (Belhaven).

Lowe, P. & A. Flynn 1989. Environmental politics and policy in the 1980s. In *The political geography of contemporary Britain*, J. Mohan (ed.), 225–79. London: Macmillan.

Lowe, P. & W. Rüdig 1986. Political ecology and the social sciences. *British Journal of Political Science* **16**, 513–50.

Lowe, P. & S. Ward (eds) 1998. *British environmental policy and Europe*. London: Routledge.

Lowe, P. & N. Ward 1997. Field-level bureaucrats and the making of new moral discourses in agri-environmental controversies. In *Globalizing Food: agrarian questions and global restructuring*, D. Goodman & M. Watts (eds). London: Routledge.

Lowe, P., G. Cox, M. MacEwen, T. O'Riordan, M. Winter 1986. *Countryside conflicts: the politics of farming, forestry and conservation*. Aldershot, England: Gower.

Lowe, P., J. Clark, S. Seymour, N. Ward 1992. *Pollution control on dairy farms: an evaluation of current policy and practice*. London: SAFE Alliance.

Lowe, P., J. Murdoch, T. Marsden, R. Munton, A. Flynn 1993. Regulating the new rural spaces: issues arising from the uneven development of land. *Journal of Rural Studies* **9**, 205–22.

Lowe, P., J. Murdoch, G. Cox 1995. A civilised retreat? Anti-urbanism, rurality and the making of Anglo-centric culture. In *Managing cities: the new urban context*, P. Healey, S. Cameron, S. Davoudi, S. Graham and A. Mandani-Pour (eds), 63–82. Chichester: John Wiley.

Lowe, P., N. Ward, S. Seymour, J. Clark 1996. Farm pollution as environmental crime. *Science as Culture* **25**, 588–612.

MacKinnon, N., J. Bryden, C. Bell, A. Fuller, M. Spearman 1991. Pluriactivity, structural change and farm household vulnerability in Western Europe. *Sociologia Ruralis* **31**, 58–71.

Maloney, W. & J. Richardson 1994. Water policy-making in England and Wales: policy communities under pressure. *Environmental Politics* **3**, 110–38.

— 1995. *Managing policy change in Britain: the politics of water*. Edinburgh: Edinburgh University Press.

Marr, A. 1996. Town vs country: a rampant urban moralism has been unleashed upon the countryside. *The Independent* (29 March).

Marsden, T., P. Lowe, S. Whatmore (eds) 1990 *Rural restructuring: Global processes and their responses*. London: David Fulton.

Marsden, T., J. Murdoch, P. Lowe, R. Munton, A. Flynn 1993. *Constructing the countryside*. London: UCL Press.

Mason, P. 1992. *Farm waste storage: guidelines for construction* [CIRIA Report 126]. London: Construction Industry Research and Information Association.

McCloughlin, J. 1975. Control of farm pollution. *Journal of Planning and Environmental Law* **4**, 78.

Milne, R. 1989. Parasite in farm waste threatens water supplies. *New Scientist* (29 July), 22.

Ministry of Agriculture, Fisheries and Food 1975. *Food from our own resources* [Cmnd 6020]. London: HMSO.

Ministry of Agriculture, Fisheries and Food 1979. Agriculture in 2000 AD. In *Agriculture and pollution* [Seventh Report of the Royal Commission on Environmental Pollution [Cmnd 7644], 242–68]. London: HMSO.

— 1980. *Slurry handling: useful facts and figures* [Booklet 2356]. London: MAFF Publications.

— 1982a. *Profitable utilization of livestock manures* [Booklet 2081, revised 1982]. London: MAFF Publications.

— 1982b. *The storage of farm manures and slurries* [Booklet 2273, revised 1982]. London: MAFF Publications.

— 1983. *General information: farm waste management* [Booklet 2077, revised 1983]. London: MAFF Publications.

— 1985. *Code of good agricultural practice* [PB0100]. London: MAFF Publications.

— 1987. Memorandum of evidence to the Committee. See House of Commons Environment Committee (1987) 114–19.

— 1992. Pilot study to cut farm waste pollution [MAFF News Release, 21 January] London: Ministry of Agriculture, Fisheries and Food.

— 1993. Changes to the Farm and Conservation Grant Scheme [MAFF News Release 418/93, 30 November] London: Ministry of Agriculture, Fisheries and Food.

— 1994a. Further action against farm waste pollution [MAFF News Release 322/94, 23 August]. London: Ministry of Agriculture, Fisheries and Food.

— 1994b. Farm waste handling grants to end [MAFF News Release 444/94, 29 November]. London: Ministry of Agriculture, Fisheries and Food.

Ministry of Agriculture, Fisheries and Food/Welsh Office Agriculture Department 1991. *Code of good agricultural practice for the protection of water*. London: MAFF Publications.

Mitchell, J. 1983. Case and situation analysis. *Sociological Review* **31**, 187–211.

Moore, N. 1987. *The bird of time: the science and politics of nature conservation*. Cambridge: Cambridge University Press.

Mumford, L. 1961. *The city in history*. London: Secker & Warburg.

Murdoch, J. 1994. *Weaving the seamless web: a consideration of network analysis and its potential application to the study of the rural economy*. Centre for Rural Economy Working Paper 3, Department of Agricultural Economics, University of Newcastle upon Tyne.

— 1995. Actor networks and the evolution of economic forms: combining description and explanation in theories of regulation, flexible specialisation and networks. *Environment and Planning A* **27**, 731–57.

— forthcoming. Inhuman/nonhuman/human: actor-network theory and the prospects for a non-dualistic and symmetrical perspective on nature and society. *Environment and Planning D: Society and Space*.

National Audit Office 1995. *National Rivers Authority: river pollution from farms in England – report by the Comptroller and Auditor General* [HC 235 Session 1994–5]. London: HMSO.

National Farmers' Union 1987. Memorandum of evidence to the Committee. See House of Commons Environment Committee (1987) 136–43.

Bibliography

National Farmers' Union 1990. *Proposed control of pollution regulation: a statement by the NFU*. London: National Farmers' Union.

— 1992. *NFU welcomes recommendations on pollution* [NFU Press Release, 21 January 1992]. London: National Farmers' Union.

National Rivers Authority 1989a. *Fact sheet 13: conservation*. London: National Rivers Authority.

— 1989b. *Guardians of the water environment* [NRA leaflet]. London: National Rivers Authority.

— 1990a. *Annual report and accounts 1989/90*. London: National Rivers Authority.

— 1990b. *Water pollution from farm waste 1989: England and Wales*. London: National Rivers Authority.

— 1991. *The quality of rivers, canals and estuaries in England and Wales: report of the 1990 survey* [Water Quality Series 4]. Bristol: National Rivers Authority.

— 1992a. *The influence of agriculture on the quality of natural waters in England and Wales*. Bristol: National Rivers Authority.

— 1992b. *Annual report and accounts – 1991/2*. Bristol: National Rivers Authority.

— 1992c. *Water pollution incidents in England and Wales – 1991*. [Water Quality Series 9] Bristol: National Rivers Authority.

— 1992d. *Water pollution incidents in England and Wales – 1990*. [Water Quality Series 7] Bristol: National Rivers Authority.

— 1993. *Water pollution incidents in England and Wales – 1992* [Water Quality Series 13]. Bristol: National Rivers Authority.

— 1994. *Water pollution incidents in England and Wales – 1993* [Water Quality Series 21]. Bristol: National Rivers Authority.

— 1995. *Water pollution incidents in England and Wales – 1994* [Water Quality Series 25]. Bristol: National Rivers Authority.

National Rivers Authority/Ministry of Agriculture, Fisheries and Food 1990. *Water pollution from farm waste 1989 (England and Wales)*. London: National Rivers Authority.

National Rivers Authority South West Region 1992. *NRA in New Year pollution blitz* [National Rivers Authority South West News Release, 8 January]. Exeter: National Rivers Authority South West.

Natural Environmental Research Council 1987. Memorandum of evidence to the Committee. See House of Commons Environment Committee (1987) 74–82.

Nature Conservancy Council 1987. Memorandum of evidence to the Committee. See House of Commons Environment Committee (1987) 256–60.

— 1991. *Nature conservation and pollution from farm wastes*. Peterborough: Nature Conservancy Council.

Newby, H., C. Bell, D. Rose, P. Saunders 1978. *Property, paternalism and power*. London: Hutchinson.

Nielson, V. 1990. Farm waste management: the challenge of the next decade. *Journal of the Royal Agricultural Society of England* **151**, 187–200.

Norgaard, R. 1994. *Development betrayed: the end of progress and a co-evolutionary revisioning of the future*. London: Routledge.

Paice, C. 1991. Your solution to pollution. *Farmers Weekly* (21 June), 62–6.

Payne, M. 1986. Agricultural pollution – the farmers' view. In *Effects of land use on fresh waters,* J. F. de L. B. Solbe (ed.), 329–34. Chichester: Ellis Horwood.

Pearce, F. 1986. A green unpleasant land. *New Scientist* (24 July), 26–7.

Porter, T. 1995. *Trust in numbers: the pursuit of objectivity in science and public life.* Princeton, New Jersey: Princeton University Press.

Revill, G. and S. Seymour 1996. Dishing the dirt: telling stories of Pollution Inspectors, ethnography and rural research. Paper presented at the RSG/IBG Conference, Glasgow.

Richardson, S. J. 1976. Proceedings of Conference on Agriculture and Water Quality *Agricultural development and advisory service technical bulletin* No. 32. London: Ministry of Agriculture and Fisheries and Food.

Richardson, J., W. Maloney, W. Rüdig 1991. *Privatising water.* Strathclyde Papers on Government and Politics 80, Department of Government, University of Strathclyde.

Royal Commission on Environmental Pollution 1971. *First report* [Cmnd 4585]. London: HMSO.

— 1974. *Pollution control: progress and problems* [Fourth Report, Cmnd 5870]. London: HMSO.

— 1979. *Agriculture and pollution* [Seventh Report, Cmnd 7644]. London: HMSO.

— 1992. *Freshwater quality* [Sixteenth Report, Cmnd 1966]. London: HMSO.

Saunders, P. 1985. The forgotten dimensions of central–local relations: theories of the "regional state". *Environment and Planning C* **3**, 149–62.

Schofield, K., J. Seager, R. Merriman 1990. The impact of intensive farming activities on river quality: the eastern Cleddau catchment study. *Journal of the Institute of Water and Environmental Management* **4**, 176–86.

Scott, J. 1976. *The moral economy of the peasant: rebellion and subsistence in Southeast Asia.* New Haven, Connecticut: Yale University Press.

— 1985. *Weapons of the weak: everyday forms of peasant resistance.* New Haven, Connecticut: Yale University Press.

— 1989. Everyday forms of resistance. In *Everyday forms of peasant resistance,* F. Colburn (ed.), 3–33. New York: M. E. Sharpe.

Seymour, S., P. Lowe, N. Ward and J. Clark 1997. Environmental "others" and "elites": rural pollution and changing power relations in the countryside. In *Revealing rural "others": representation power and identity in the British countryside,* P. Milbourne (ed.), 57–74. London: Pinter.

South West Water Authority 1986. *Environmental investigation of the River Torridge.* Exeter: Department of Environmental Services, South West Water Authority.

— 1987. Memorandum of evidence to the Committee. See House of Commons Environment Committee (1987) 306–9.

Storey, D. 1977. A socio-economic approach to water pollution law enforcement in England and Wales. *International Journal of Social Economics* **4**, 207–24.

Thompson, P. 1995. *The spirit of the soil: agriculture and environmental ethics.* London: Routledge.

Trevelyan, G. 1942. *English Social History.* Harlow, England: Longman.

Vogel, D. 1986. *National styles of regulation: environmental policy in Great Britain and the United States*. Ithaca, New York: Cornell University Press.

Ward, N. 1993. The agricultural treadmill and the rural environment in the post-productivist era. *Sociologia Ruralis* **33**, 348–64.

— 1995. Technological change and the regulation of pollution from agricultural pesticides. *Geoforum* **26**, 19–33.

Ward, N. & P. Lowe 1994. Shifting values in agriculture: the farm family and pollution regulation. *Journal of Rural Studies* **10**, 173–84.

Ward, N., P. Lowe, S. Seymour, J. Clark 1995a. Rural restructuring and the regulation of farm pollution. *Environment and Planning A* **27**, 1193–211.

Ward, N., H. Buller, P. Lowe 1995b. *Implementing European environmental policy at the local level: the UK experience with water quality directives* [2 volumes]. Research Report, Centre for Rural Economy, University of Newcastle upon Tyne.

— 1996. The Europeanisation of local environmental politics: bathing water pollution in south west England. *Local Environment* **1**, 21–32.

Watchman, P., C. Barker, J. Rowan-Robinson 1988. River pollution: a case for a pragmatic approach to enforcement. *Journal of Planning and Environmental Law* **17**, 674–9.

Water Authorities Association 1987. Memorandum of evidence to the Committee. See House of Commons Environment Committee (1987) 25–8.

— 1986. *Water pollution from farm waste 1985 (England and Wales)*. London: Water Authorities Association.

Water Authorities Association/Ministry of Agriculture, Fisheries and Food 1986. *Water pollution from farm waste 1985 (England and Wales)*. London: Water Authorities Association.

— 1987. *Water pollution from farm waste 1986 (England and Wales)*. London: Water Authorities Association.

— 1988. *Water pollution from farm waste 1987 (England and Wales)*. London: Water Authorities Association.

— 1989. *Water pollution from farm waste 1988 (England and Wales)*. London: Water Authorities Association.

Welsh Water Authority 1984. Memorandum of evidence to the Committee. In *Agriculture and the environment*, House of Lords Select Committee on the European Communities, 348–52. London: HMSO.

Weller, J. & S. Willetts 1977. *Farm wastes management*. London: Crosby Lockwood Staples.

Whatmore, S. 1994. Global agro-food complexes and the refashioning of rural Europe. In *Globalization, institutions and regional development in Europe*, N. Thrift & A. Amin (eds), 46–67. Oxford: Oxford University Press.

Wilkinson, M. 1990. *Silage UK*, 6th edn. Marlow, Buckinghamshire: Chalcombe.

Williams, R. 1976. *Keywords: a vocabulary of culture and society*. London: Fontana.

Wilmot, S. 1993 Agriculture and pollution in Victorian Britain. Paper presented at the combined British Agricultural History Society and Institute of British Geographers annual conference, Institute of Historical Research, London.

Index

salmon 75
Salmon Fisheries Act 1861 40
Salmon and Fresh Water Fish Act 1923
 40
Samaritans 190
Sandford, Lord 48
Saunders, P. 63
Schofield, K. 37
scientific expertise 9, 50, 195–7
Scott, J. 200, 205
Secretary of State for the Environment
 47–9
self-regulation, *see* voluntarism
Severn Trent Water Authority 67, 72
sewage 7, 24, 35–6, 39, 40, 42, 54,
 69–70, 74–7, 89
silage 6, 24, 28–34, 40, 45, 50–1,
 67–8, 76, 122, 131, 179
Sites of Special Scientific Interest 70
slurry 20, 24, 33–4, 50–2
 slurry spreading 19–20, 29, 34–5,
 50–3, 70
 slurry storage 21–34, 82, 120–1,
 134, 146–54, 197–8
social change in the countryside 3,
 13–14, 173–82
social construction of pollution 4, 6,
 9–12, 183–7, 208
social regulation 2, 8
sociology of translation 9–12
 see also enrolment; actor network
 theory
South West Water Authority 6, 64, 67,
 75–7, 79–81, 90–2
stewardship 125–7, 143–4, 201–2
stocking rates 20, 45, 50, 52
Storey, D. 42
street level bureaucrats 203–7
Strontium-90 3
Strutt, Sir Nigel 42–6
suicide 190
symmetry 11

technical advice 3, 146–54
technical discourse of farm pollution 1,
 44–6, 50–3, 57–9, 127–36, 146–54,
 162–73
technical solutions 49, 53, 146–54,
 170

technological change 1, 4, 18–37,
 201–2
Thatcher, M. 61
Thompson, P. 201–2
Thoms, R. 68
town dairies 7, 95
Townsend, A. 154
Trevelyan, G. M. 2

urban-rural dualism 2, 208
 see also rurality

Vogel, D. 32
voluntarism 8, 27, 78, 86, 187

Ward, N. 39, 65, 174, 201
Ward, S. 189
Watchman, P. 104
Water Act 1973 49
 see also water privatization
Water Act 1989 27, 88, 115–16
water authorities 40, 46–8, 54–7,
 62–70, 78–88, 166, 171, 184, 196
Water Authorities Association 62–8,
 70–4
water companies 89
Water Companies' Association 71
water engineers 63
water industry 49, 54, 61–3, 70–1, 82
water policy community 61–3
water protection zones 47–8
water privatization 15, 39, 61–3, 83, 155
 see also Water Act; water industry
water quality 37, 53, 69, 70, 71
 see also pollution; biochemical oxygen
 demand
Water Research Centre 71
weapons of the weak 205–6
Welsh Water Authority 67, 70
Weller, J. 40, 42–3, 55
Welsh Office 37, 61, 68, 155, 165
Wessex Water 68
Western Morning News 9, 155, 157–61
White Paper on agriculture 49
Wildlife and Countryside Act 1981 86
Wilkinson, M. 30
Willetts, S. 40, 42–3, 55
Wilmot, S. 7
Working Party on Sewage Disposal 43